129669

Springer Texts in Education

More information about this series at http://www.springer.com/series/13812

Jürgen Maaß · Niamh O'Meara
Patrick Johnson · John O'Donoghue

Mathematical Modelling for Teachers

A Practical Guide to Applicable Mathematics Education

Springer

Jürgen Maaß
Institut für Didaktik der Mathematik
Universität Linz
Linz, Austria

Patrick Johnson
School of Education
University of Limerick
Limerick, Ireland

Niamh O'Meara
EPISTEM, National Centre for STEM Education, School of Education
University of Limerick
Limerick, Ireland

John O'Donoghue
EPISTEM, National Centre for STEM Education, School of Education
University of Limerick
Limerick, Ireland

ISSN 2366-7672 ISSN 2366-7680 (electronic)
Springer Texts in Education
ISBN 978-3-030-00430-9 ISBN 978-3-030-00431-6 (eBook)
https://doi.org/10.1007/978-3-030-00431-6

Library of Congress Control Number: 2018954031

This Springer imprint is published by the registered company Springer Nature Switzerland AG
The registered company address is: Gewerbestrasse 11, 6330 Cham, Switzerland

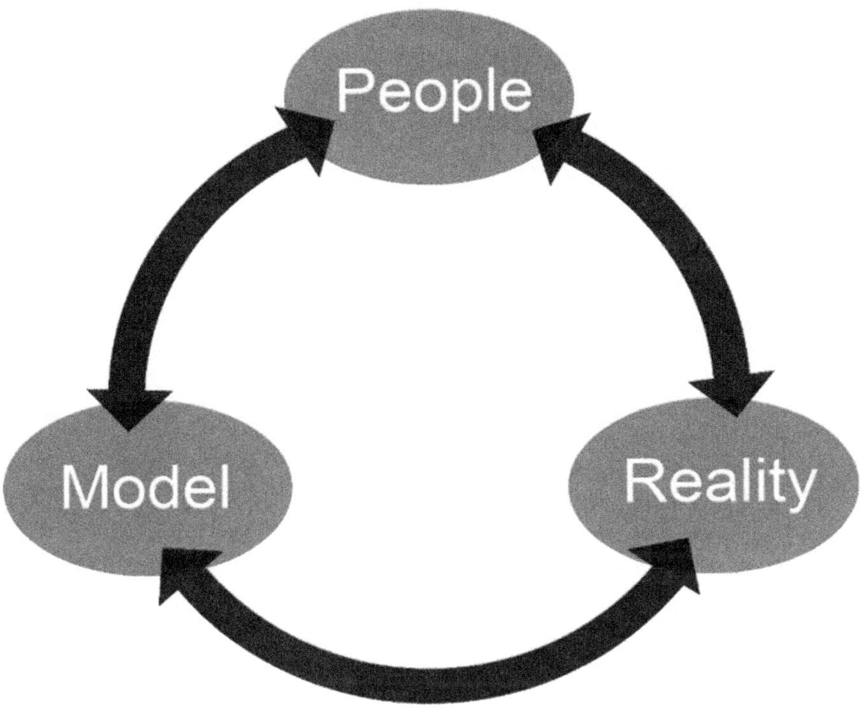
People
Model
Reality

Preface

While there are many areas of focus in mathematics education, there are many good reasons for offering *applicable* mathematics education in schools. Let us just mention two of the most important reasons. On the one hand, a focus on the practical side of mathematics presents a convincing and motivating answer to the typical student question: 'Why study mathematics?' On the other hand, education policy seems inclined to move in this direction by implementing international testing, curricula and catalogues of skills.

Here, *applicable mathematics education* is the phrase we use to describe reality-based mathematics education. Reality-based mathematics relies heavily on problem-solving and a positive disposition to engage with mathematics. Modelling reality and simulating selected aspects of reality are other pillars of reality-based mathematics education.

One of the insights into recent progress of new and reform mathematics curricula, which is in line with international trends, is to learn how crucial it is to convince teachers that it is worthwhile to conduct applicable mathematics classes and also to develop their ability to do so. How can we manage to achieve this in the training and continuing education of mathematics teachers? This book, a short study aid for teachers, is part of the answer. It is a book to help you, the mathematics teacher, understand, participate, experiment and be motivated about your discipline.

The most important feature of the book is that it tries to speak directly to mathematics teachers. The authors draw teachers into a continuous dialogue about activities they are asked to engage in themselves as learners. They are asked to do something, and through doing and reflection, they gain first-hand experience of new approaches and materials. In this way, they can learn to teach applicable mathematics to their students using their own experience as learners of applicable mathematics, motivated and supported by the book. The book also shows teachers how to extend tasks/examples by offering hints for consideration and identifying further accessible resources. Finally, the authors present some background information on the modelling process and related ideas, and the professional applications of mathematics. At this point, we should highlight that this work is based on the lead author's (J. Maaß) original book[1] published in the German language.

As regards supports for the implementation of applicable mathematics education, mathematics teachers in German-speaking countries have been very well served for several decades by two networks MUED (www.mued.de) and ISTRON (http://istron.uni-koblenz.de/istron/). The situation in English-language countries is less organised and coherent when it comes to dedicated supports for reality-based mathematics education, but they do exist and can be accessed by teachers. Generally, one needs to look at this country by country as there are few, if any, dedicated pan-national networks or associations. However, many national mathematical organisations and associations provide reality-based materials for use in schools and are accessible by teachers everywhere through websites and published materials. Here, we also recognise the role of international conferences such as CERME, ICME, ICTMA and others that host themed sessions and topic study groups on problem-solving, modelling and applications of mathematics. The authors direct mathematics teachers to appropriate sources and resources as they progress through the book.

Of course, one book cannot guarantee all our wishes are fulfilled regarding the improvement of mathematics education. Nevertheless, it can achieve much more than merely keeping one informed of some developments in the subject. This book is different in that regard because it deviates from the standard set-up in mathematical studies by simultaneously implementing didactical insights into its structure, writing style and in its dealing with its content. This approach is not customary in books for teachers, as was recognised many years ago by Heinz Hülsmann who criticised the anonymous style of didactics in writing about itself, and the situation has not changed. However, in this book it is apparent that didactics is not merely being described, but applied in a hands-on fashion to explain its content.

When we talk about 'didactics', we are specifically referring to 'didactics of mathematics' as understood in the German, Austrian, French, Dutch, Scandinavian and other traditions, where the 'didactics' of a school subject, in this case mathematics, is closely associated with its parent discipline, namely mathematics. The central concerns of the didactics of mathematics are the contents of school mathematics; teaching; learners; the mathematics curriculum; the preparation of teachers and matters relating to the practice of teaching school mathematics. In recent years, some have interpreted the concerns of the didactics of mathematics as moving outside of the institutional setting of schools into the area of vocational education, workplace mathematics and adult mathematics education.

The terms pedagogy and didactics are often used interchangeably, as in *mathematics pedagogy* and *didactics of mathematics*, but we take the view that pedagogy is more education and learner-focused, while didactics is discipline and

[1] Maaß, J. (2015). *Modeling in school. A learning book on theory and practice of application-oriented mathematics teaching. (Modellieren in der Schule. Ein Lernbuch zu Theorie und Praxis des realitätsbezogenen Mathematikunterrichts: Schriften zum Modellieren und zum Anwenden von Mathematik 5).* Münster: WTM-Verlag.

teacher-focused. We acknowledge the reality in mathematics education that there is considerable overlap between the two, and a hard and fast distinction is difficult to maintain.

These technical matters should not detract from the usefulness of the subject matter of the book for a wider audience of practicing mathematics teachers, mathematics educators and researchers around the world.

Linz, Austria — Jürgen Maaß
Limerick, Ireland — Niamh O'Meara
Limerick, Ireland — Patrick Johnson
Limerick, Ireland — John O'Donoghue

Contents

About the Authors

Jürgen Maaß (Juergen Maasz in English) is a Professor for Mathematics Education at the Kepler University in Linz (Austria). He started thinking about improving teaching mathematics by using real-world problems about 40 years ago when he joined the group 'MUED' (see: www.mued.de) a year before it was founded officially by Heinz Böer. He is a prominent member, author and editor of several books of the 'ISTRON' group (www.istron.mathematik.uni-wuerzburg.de/).

Other areas of his research and published books and papers are mathematics and society (mathematics as technology, mathematics as a social system, industrial mathematics, adults and mathematics), mathematics and philosophy (ethics, theory of recognition). He was previously the chair of Adults Learning Mathematics (ALM) and of the Austrian Mathematics Educators.

Niamh O'Meara is a Lecturer in Mathematics Education in EPISTEM, the National Centre for STEM Education, at the University of Limerick. She has worked as a lecturer since 2014 and prior to that was the Senior Project Officer (Mathematics) in the National Centre for Excellence in Mathematics and Science Teaching and Learning (NCE-MSTL). She was awarded her Ph.D. in 2011 from the University of Limerick. This work investigated the knowledge required to teach mathematics effectively, and she developed a model of teacher knowledge for the twenty-first century. Her current research interests include teacher knowledge; mathematics teacher education; numeracy; mathematics in the workplace and issues surrounding the mathematics curriculum.

Patrick Johnson is a Lecturer in Mathematics Education in the School of Education at the University of Limerick where he has been a faculty member since 2012. Prior to this, he was a Research Fellow in the National Centre for Excellence in Mathematics and Science Teaching and Learning. He received a Doctor of Philosophy (Ph.D.) from the University of Limerick in the field of Applied

Mathematics before switching in to the area of mathematics education. His current research interests focus on the topics of mathematical modelling, problem-solving and attitudes towards mathematics.

John O'Donoghue is an Associate Professor (Emeritus) of Mathematics Education at the University of Limerick. He has served as the Co-Founder and Director of the National Centre for Excellence in Mathematics and Science Teaching and Learning (NCE-MSTL), now EPI*STEM: National Centre for STEM Education, at the University of Limerick (UL). He is the Founder and past Director of the Mathematics Learning Centre, UL. He is also a past chair and honorary trustee of *Adult Learning Mathematics—A Research Forum*. He has a long history of involvement with mathematics teacher education in Ireland. His research interests include mathematics teacher education (secondary), adult mathematics education/numeracy, service mathematics teaching and mathematics learner support (HE). He has supervised many Ph.D. and master's research candidates to completion.

Chapter 1
Introduction to Modelling

Some questions that are likely to arise for mathematics teachers who are considering implementing reality-based mathematics are: What is modelling in mathematics education? Why should I as a mathematics teacher engage with applicable mathematics in my classroom? How do I go about this? What can I do if my students are not able to apply their mathematics or do not want to try? How do I get started? What examples could I use to develop more extensive course units? Where can I find more examples?

The most important feature of this book is that is tries to speak directly to you, the mathematics teachers. The authors attempt to draw you into a continuous dialogue about activities you are asked to engage in as learners. You are asked to do something, and through doing and reflecting you will gain first-hand experience of new approaches and materials. In this way, you can learn to teach applicable mathematics to your students using your own experience as learners of applicable mathematics, motivated and supported by the book.

Here *applicable mathematics education* is the phrase we use to describe reality-based mathematics education. Reality-based mathematics relies heavily on problem solving and a positive disposition to engage with mathematics. Modelling reality and simulating selected aspects of reality are other pillars of reality-based mathematics education.

Let us start with a short description of what is known, in the didactics of mathematics, as the modelling cycle. It is a process that starts with the desire to be able to better understand something or some situation, or to change it. We proceed by selecting the relevant information and identifying our goals, then applying mathematics to our initial situation and repeating the process several times until we achieve the desired result, or we realise, based on the available evidence, that it will not work this way. There is a large and expanding literature on problem solving and modelling and several diagrams are used to depict the modelling process (Blum et al. 2007; Schoenfeld 1992). In the last chapter of this book you will find a discussion and more details on the modelling cycle.

At the outset there is a decision—there must be a decision to engage. The desire to realise, better understand, change or achieve something faster, to control with less

J. Maaß et al., *Mathematical Modelling for Teachers*, Springer Texts in Education, https://doi.org/10.1007/978-3-030-00431-6_1

effort or to put resources to more efficient use can serve as a motivating force in arriving at a decision. First, we select one aspect of reality that we want to, and are able to, look at in more detail by employing mathematical methods, and then we search for data and laws that correlate with our data.

In the beginning we often lack necessary data or it is not apparent how the available data should be structured using our mathematical toolkit or how it can be presented and described as a mathematical equation. Therefore, we frequently start by making approximations and/or treating the data in very simple mathematical terms (i.e. mathematizing). Initially, we create models (that may involve mathematics) which clearly do not take into account all aspects of the problem or situation. Nevertheless, we soon set out to do calculations that allow us to interpret the result as well as the path to how we got there. This interpretation will take us in different directions. The initial results often cause us to question our progress, as the results may not be satisfactory or we get the impression that the problem does not warrant any further effort. Often we are not satisfied with the results and by returning and analysing our first attempt, we can then decide how to continue.

Resulting problems often require more detailed and additional data, more complex mathematical tools or a more precise description of the actual problem or goals. As soon as we have considered these issues, we make a second attempt, and the outcome has to be analysed once again. Similar to the first cycle we now have to consider whether and how to proceed.

After a few attempts our findings will tell us, in sufficient detail, what we wanted to know. Otherwise, we realise that we are stuck despite great effort, because better data is not available; the mathematical representation is beyond our cognitive skills; or we come to the conclusion that pursuing the matter further would require too much effort. However, mathematicians in the private sector have to deal with this issue on a daily basis and it shows us that collecting relevant data is often an arduous task. In the next section we will take a closer look at this and invite you to engage with your first (simple) modelling problem with us, which involves looking for a current advertisement for a mobile phone contract and checking if it really is as cheap as the advertisement claims. Let's begin.

1.1 Mathematical Modelling of Real-Life Mobile Phone Tariffs

We assume everyone aims to use their mobile phone as cost-efficiently as possible. Several examples on this subject have been previously published (Cheng and Chua (2015) or http://dpssecondarymathematics.pbworks.com/w/file/51453580/MIRL_Linear_Functions_1.pdf).

We would like to present this topic in an interactive way. In this case many steps have already been predefined. However, at the start of a mathematical modelling exercise such ideas usually have not been generated yet. Let us take a closer look at this during our first task.

> Please take a closer look at a price plan from your service provider. If you are struggling to find one, please check out the following website for an example (see p. 91)
> http://www.productsandservices.bt.com/products/lib/pdf/BT_PhoneTariff_Residential.pdf?s_cid=con_FURL_tariffguide

Having accessed the website, are you like us, surprised by the flood of information and details on this website? If this is the case then you might also need a filter for this excess information, just like we did, and so will your students later on. The goal is to identify as few as possible distinct factors that would allow us to sort this information into two categories: important and not important.

Here we insist that the precondition for successful modelling is a clear definition of its goals. We assume that you will agree with us on this point, at least for now. The less defined the goals are, the harder it is to identify the information necessary to attain them. Clearly defined goals help us settle all decisive questions that arise along the way. For example, should this detail also be taken into account? Should we not also pay attention to some other issue?

As regards goals, at this point you might want to consider the proposition that research always attempts to find something previously unknown (or to make something already known more accessible for a certain purpose or goal). Therefore, it cannot always be determined in advance what the goal will be. However, if the goal is known, half the work is already done. This is almost always true when the goal is known!

Research in general, and mathematical modelling in particular, has one characteristic that often goes unmentioned when success is being reported afterwards; it advances step-by-step with a preliminary goal leading to the first step. Lessons can, and should, be learned from successes and failures encountered on the initial course of action taken. An integral part of the learning process is the opportunity to make more precise statements about the goals based on knowledge gained so far. In this way, especially where technological advances are concerned, time and again we encounter limitations that cannot be exceeded, such as laws of nature or properties of materials.

> In our virtual mobile phone project you have to define your own goals! The following list of indicative questions will help:
> Would you like to spend as little as possible on your phone bill?
> Do you want talk time (calls) to be the main influence on your bill?
> Would you like to use additional services?

> Would you like to be constantly available and talk on your phone all the time, or do you just need a means of communication in case of emergency?
> Have you reflected a little on your goals? Please write down your thoughts.
> Later compare your notes with your (preliminary) result(s).

Let us now proceed to our second task:

> Please think about what information you need to collect to be able to compare the advertised tariff with your current one.
> Please write down your ideas.

At first it might seem tedious to write down the caption 'Information on my current mobile usage', but this active participation in the first example in this book should show you, and similarly your students, how an abstract outline of modelling can be done using a real-life example. Your notes from the very beginning will prove useful for understanding what has been achieved so far by employing these methods, and what can be gained from taking similar notes in future modelling exercises.

Let us return to the task in hand. Have you found your recent mobile phone bill? In addition to the total amount due you might find additional details such as the types of phone numbers called (premium, landline, mobile, freephone), total number of calls made, internet usage or the number of text messages sent. Now we are getting to the harder part: have you been using your mobile phone the same way in the past month as in previous months? Or in other words, was this bill for a month during which all the services you used incurred average costs?

To better analyse the way you have been using your mobile phone so far, it makes sense to survey your use over a longer period of time. How far back should you go with your mobile phone bills? This is up to you! Would you like to focus on your everyday use? Should the answer be yes, then bills issued in the last few months are enough. Would you also like to make calls very cheaply during your vacation or play games or follow your favourite sports or watch TV on the internet? In this case it helps to specify your goals. Please return to your notes from before! Were your goals detailed enough? If the answer is 'yes' then you already have your criteria. In case of a 'no' answer you will probably better understand now why we asked you to consider your goals in more detail, and redefine them now for your mobile phone project.

The next step is to consider some more ideas: whether you use your mobile phone with the same provider or a different one, you might have reasons for changing the way you use your mobile. This is not easy to predict, you might say. Now you have identified a typical challenge in modelling—the future is not easy to predict. Data about the past however is fairly easy to compile, even if it takes a little effort sometimes. However, data about the future is not precise, otherwise gambling would not exist.

If you do want to decide on a particular tariff for your mobile phone contract by making assumptions that take account of future uncertainties about tariffs, you will need to show a little creativity in determining possible (future) user behaviour.

Please also make notes on this. For example, if I got a new smart phone, I would like to use the internet, especially for emails.

Since you have been actively following us so far, you now know more about yourself as a mobile phone user, your requirements for a suitable mobile phone tariff and your expected future user behaviour. Optimistically, attempts at mathematizing frequently results in only one aspect of everyday behaviour being analysed—and maybe this analysis will influence our rational conduct or our decisions. We shall return to this later.

Maybe you have been asking yourself for quite a while now what all this has got to do with mathematics? I have not calculated anything yet. There is much more to mathematics than doing calculations. The book by Davis et al. (2012), entitled *The Mathematical Experience, Study Edition*, underlines this viewpoint. Also many authors would agree that mathematical modelling and solving real world problems are also well established parts of mathematics. Every realistic project requires a predefinition of its goals as well as searching, compiling and analysing data, and modelling. As previously mentioned, this may not happen on a once-off basis. It is a work in progress that will hopefully, step-by-step, take us closer to our goal, using the data already compiled and interpreting it in line with our predefined goals. All of these processes are an intrinsic part of mathematics as well and are mathematical processes. Developing these skills also aligns with the requirements of school curricula.

1.2 Decisions at the Outset

At the start, working with all the information on mobile phone use is rather complicated. We therefore recommend selecting the service used most frequently, such as the cost of phone calls to other mobile phone numbers.

Here we invite you to list the pros and cons for such a rigorous pre-selection.

The main counter-argument is obvious: you are using various services on your mobile! Should you consider changing your contract, it only makes sense if the invoice total decreases (provided network coverage will be at least as good as it is now). What good is it if the cost of phone calls decreases but the cost of texts or other services increases considerably?

The main argument in favour of choosing the service you use most frequently is of a rather pedagogical nature: it probably makes things much easier at the start of your modelling experience—and certainly that of your students. Should you prefer to start with a higher level of complexity, you are welcome to do so. We will catch up with you after a few steps of the modelling process in the later section—*Thoughts for Teachers on our First Model*.

Before we begin modelling, we would like to emphasise how rare it is for students to have an overview of their mathematical studies, to plan their calculations and decide which steps to take before actually starting to calculate something. Frequently, teachers complain about this, but they rarely let their students gain experience in doing such activities. Often the teacher enters class with a predefined and prepared plan for the classroom lesson; they explain new material, repeat algorithms already studied or instruct students to solve a selected problem from the textbook. Discussions about the curriculum, the sequencing of topics to be covered, or the selection of textbook problems to be solved are not encouraged, due in no small part to time constraints and the perception that such discussions are beyond the capabilities of students.

We get an entirely different picture in an applicable mathematics class, when the first model has been set up and decisions have to be made about it. Why is the situation different in your regular mathematics class? It is paramount that the decisions have an immediate effect on student activity. The class considers and decides together what the most sensible thing to do next is. The decisions taken have an immediate effect on students, rather than some unrelated judgements taken by other people, for example, some curriculum committee or government department concerned with deliberations on curricula. If the topic of instruction has been selected in such a way so that students can apply their own life experience—as is the case in our mobile phone tariffs example—they should get the opportunity to do this (Stillman et al. 2013). If they learn to act based on their own decisions in the course of such lessons, mathematics education has also contributed to fostering a range of life skills. Naturally, teachers can, and are supposed to, support their students in this. This support may take the shape of advising them about available options (Is there an algorithm for this? Is there an equation for this?), as well as decision-making procedures (a simple vote is not always the best option).

1.3 A First Modelling Exercise

Take a look at your mobile phone bill, or the BT advertisement previously provided, or some other available information on tariffs.
What comes to mind from a mathematical point of view?

We notice three types of tariffs for making calls, a flat rate (the invoice total is the same every month regardless of the number of calls made, for example, GBP 25); a tariff of GBP 10 which includes 200 min of free calls and a fixed cost per call after that (35 pence/min); or a tariff of GBP 15 which includes 500 min of free calls and a fixed cost per call after that (35 pence/min).

From a mathematician's point of view the first tariff, described above, results in a linear function while the latter two yield piecewise linear functions. Should you have come to the same conclusion, you will also think of graphing the function and analysing the graph or solving linear equations. If your students have just practiced graphing tables of values, they might realise it themselves without needing much help. Here we shall provide you with some suggestions for students who might need more assistance from you.

Your students might not be used to filtering the information and data needed from rather extensive texts on tariffs, phone bills or advertisements. Word problems in textbooks often use very precise language already pointing students in the desired direction. That way it is easy for students to 'translate' the information provided into mathematical notation, formulae or equations. Information this detailed and precise only exists in textbooks, reality is generally less specific and messy. Since your students are supposed to study mathematics for real-life situations, they need to learn how to filter the important information for themselves from less specific everyday problems. The question of what one wants to know takes centre stage in this process.

Returning to our earlier problem regarding mobile phone tariffs, please now note the information you will be looking for when you read the price plan or website. After deciding to focus on determining the costs of calls, we look precisely for this information. In the course of this search we take note of other things that may be important for refining our model at a later stage.

What information regarding the cost of calls could you gather from the BT ad provided? We have come across three different tariffs:

Plan A—Flat rate 25 GBP/month;
Plan B—10 GBP/month plan with 200 free minutes and 35 pence/min out of plan;
Plan C—15 GBP/month plan with 500 free minutes and 35 pence/min out of plan.

The following is a list of potential areas that we could focus on when we look to refine the model at a later time:

- Timing (rounding off policy)
- Fixed costs when first entering the contract (activation, importing your old number—cost depending on your provider)
- New mobile phone or not?
- Contract period?
- Network coverage?
- Cost of other services (for example text messages, data).

Let us save this list for later.

Using the information provided in relation to the three different tariff plans, you will of course be able to deduce mathematical correlations immediately. However, if your students need help with mathematizing at this stage, we suggest creating a

Table 1.1 Costs related to individual tariff plans

Hours	Plan A	Plan B	Plan C
0	25	10	15
1	25	10	15
2	25	10	15
3	25	10	15
4	25	24	15
5	25	45	15
6	25	66	15
7	25	87	15
8	25	108	15
9	25	129	29
10	25	150	50
11	25	171	71

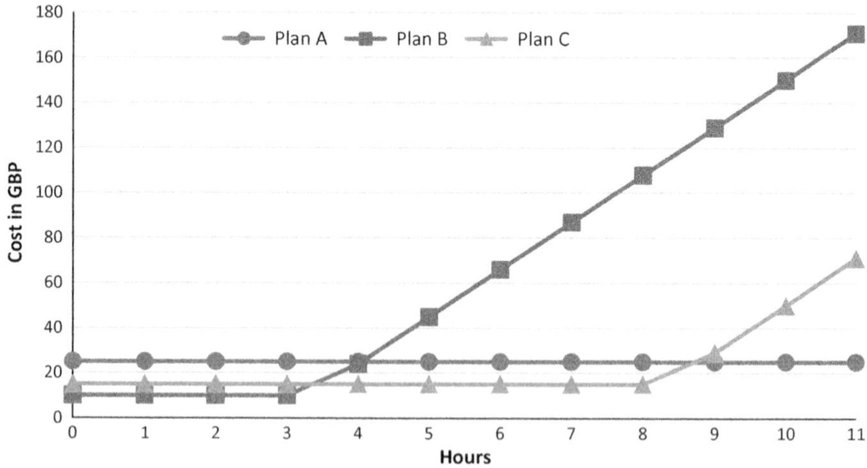

Fig. 1.1 Graph of total cost related to individual tariff plans

table of values and a chart. Putting your students in small groups (4 at the most) is very helpful here from a pedagogical perspective. In the course of the group project you will be able to tell which questions or problems arise and what already works well. In these groups students should be required to calculate the monthly charges for the three tariffs assuming a talk time of 1, 2, 3, 4… up to 11 h/month, insert the values into the table and create a chart based on this information.

We used spreadsheet analysis and created the Table 1.1 of values.

If you create a graph using this information, it should look similar to this one (Fig. 1.1).

What's next? As a mathematics teacher, based on the graph, you can immediately tell which tariff is the cheapest when we focus solely on the duration of phone calls made. But how will your students manage? If you want to provide them with some assistance, we suggest putting them to work in small groups and have them address the following questions:

1. What is the purpose of generating a table of values and a graph in this modelling exercise?
2. Do the table and the graph provide us with the information we want to know?
3. Which plan (A, B or C) would you recommend, based on the length of time that is spent on phone calls over the course of a month?
4. How precise do we have to be in stating the talk time per month? Does the table or graph provide us with precise enough data? If your answer is no, how can we obtain more precise values?

Please write some remarks on the problems posed above and the suggested responses provided below.

Question 1: It is apparent that a table of values or a graph like this can be used as a source of advice. Looking at the graph, you can easily tell which plan is best for someone who only talks on his/her mobile phone for a given number of hours per month. If the aim is only to tell which plan is best for a single case (i.e. a user with a certain number of hours talk time per month), maybe the mathematical investment is already too high. On the other hand, a certain average talk time is not a definite unit of measurement. You only arrived at this value by taking the average of the previous few monthly bills. In addition, the user might also have changed his user behaviour as a result of the new information on his tariff. For example, consider users who frequently talk on their mobiles and think they should cut back on their talk time. Upon receiving this information, they might come to the conclusion that their total bill is close to the flat rate. In this case switching to the flat rate makes talking on your mobile possible without any recriminations or ramifications such as an increase in the cost of your monthly bill. If this outcome leads to a discussion on how modelling a real-life topic, such as mobile phone tariffs, can prompt people to change their behaviour then mathematics lessons have managed to connect with reality and people's everyday lives. We shall return to this later.

Question 2: Insert the amount due (in GBP) on the cost-axis. It is important to select the function with the lowest cost for a defined talk time per month.

Question 3: The chart tells us that Plan B would be the best option up to 3.57 h/month, followed by Plan C from approximately 3.57–8.81 h, and above that Plan A (flat rate).

Question 4: In the past, the degree of precision of a result was of no great importance in mathematics education in secondary schools. Without the assistance of electronic calculators it was too burdensome to calculate something with a high degree of precision. At this point we would like to remind you of how long others (Johannes Kepler) had to work on tables of values for trigonometric functions and logarithms so as to achieve a higher degree of precision (especially for astronomical computations). With the advent of electronic aids, such as the calculator, it suddenly became very easy to calculate a result up to a required degree of precision after the decimal point. This prompted a discussion on how many decimal places after the decimal point made sense in mathematics education. In many cases however, it was decided that for all types of calculations two decimal places after the decimal point would suffice.

Should you wish for more problems for your students dealing with the tariffs, you can give your students the following tasks:

1. Please calculate how much various providers would charge you for your own user behaviour;
2. Think of a virtual person and define his/her user behaviour. You might choose someone you know. You can deliberately choose a rather unusual user such as a travelling salesman who makes lots of calls to his/her clients or a patient in hospital who has to rest in bed for a few weeks and now can make lots of calls to all of his/her friends. Now let your students calculate the cost incurred by this virtual person. Have them compare their results and ask your students to give (good) advice to their virtual person.

By the way, in coming up with this virtual person, you have also set up a mathematical model yourself. Have you also thought about his/her CV, or how this person's living room is decorated or other details about him/her that have nothing to do with his/her user behaviour? Probably not. In other words, you focused on what is important when you set up this model. You were perfectly right in doing so, as this is the way to go about modelling. What is important is determined in line with the aim of our project. This is also what makes this type of mathematics class different from conventional mathematics education, where everything is already predetermined (by the textbook or the teacher).

1.4 Thoughts for Teachers on Our First Model

Now we have to make a decision on our mobile phone contract: Should we stay with our provider or change to a new one? How will you decide, bearing in mind your (preliminary) findings? The project is finished if you do not want to switch, even if many mathematical questions still remain unanswered. How do you feel about such an outcome? After the first run through you know what you wanted to know and decide to quit, even if further considerations could reveal some interesting mathematical challenges. During our mathematical studies at university we got plenty of practice in problem solving and modelling and thus realised that a problem is only solved properly, and a theory is only understood, after we have explored further considerations.

When dealing with real-world problems, criteria other than maths-intrinsic systems or completeness apply, such as reality. Just as the problem evolved from real-life considerations, the requirements of reality determine to what extent we are going to deal with the problem and the area of mathematics it requires. As soon as we know what we wanted to know with the appropriate degree of precision, we are done, regardless of further unused options for mathematizing. What exactly is the appropriate degree of precision? Frequently, this is a compromise between attaining one's goals within a reasonable time limit using the mathematical methods at one's disposal, and precision of the real-world data available to you.

In the mathematics classroom we can put our quest for a suitable answer to the real world on hold and instead focus more on the mathematical dimension of the problem. We will demonstrate this using the topic of timing arrangements after our next section, the preliminary assessment.

1.5 Preliminary Assessment After Your First Modelling Exercise

Please review your notes on your procedural steps and write down what you have achieved so far and what you still wish to achieve.

Initially we have made a seemingly random decision so that we could get started. We focused on the total cost of calls made in a particular month. This decision helped us filter the information we needed to set up our first model from all the available information on mobile phone contracts. We found three types of tariffs and compared them using a table of values and a graph. This enabled us to identify which tariff, with a certain degree of inaccuracy, is best for which talk time.

How should we proceed from here? As previously mentioned it all depends on what we want to achieve. If a glance at our mobile phone bill tells us that we make calls for at least an hour a day, the obvious choice would be to switch to the flat rate (Plan A), provided data or texting would not be too expensive. Users who rarely make calls do not have to think long about this either; Plan B would be the obvious choice in their case.

In order to make a suitable offer for all those who do not fall into either of the simplistic cases outlined above, it is apparent that we need to undo all the simplifications we initially made. We are looking for a method to compare the cost of all types of use. Before doing so, we (and those of you who choose to work with us) will consider which factors influence the cumulative time spent on calls. A key consideration here would arise when one exceeds their free minutes of talk time and must pay for all subsequent calls.

1.6 Improving Your First Model: Details of Timing

Please gather some information from mobile phone companies regarding their rounding off policy.
What exactly is it? Why is it important?

Once a person who subscribes to Plan B or C exceeds their allocated free minutes of talk time, consideration must be given to the issue of rounding off (both in terms of duration and charge). In the BT ad provided previously in this chapter, the rounding off policy is as follows:

- Call durations are rounded up to the next whole second (with the minimum duration being 1 s).
- There is a minimum charge of 1 min.
- Call charges are rounded down to the next tenth of a penny.

We can readily agree with the following: For the consumer it is best to bill as accurately as possible—therefore select timing in minutes, as in the BT policy. If your provider uses timing arrangements of 5 min, for example, you at least have to pay for the first 5 min, even if your call only lasts a few seconds. The most common timing arrangements are minute-based.

At this point we know, for people who talk a lot on their phones, timing arrangements make a flat rate more desirable as many short calls would make the other tariff options more expensive in the long run. What can we do in class to get our students to realise this? At first students should understand the importance of when the rounding occurs. To this end, they can collect data on their own calls and analyse it. To do this, ask students to record the duration of the next ten cell calls they make as precisely as possible. For now, let us assume the following data is available to us.

Table 1.2 Sample data for call times

Call	Duration
1	1 min 14 s
2	34 s
3	3 min 54 s
4	2 min 34 s
5	45 s
6	8 min 22 s
7	5 min 45 s
8	38 s
9	1 min 12 s
10	6 min 19 s

Some questions that might immediately arise are:

1. How long did the phone calls last?
2. Hence, how many call minutes will be billed based on the rounding policy of the company?

Answers:

1. $74+35+234+154+45+502+345+38+72+379=1878$
2. The answer to this question really depends on when the company rounds your calls to the nearest minute as highlighted below:
   1878 s equal 31 min and 18 s. If rounding occurs at the end of the month, 32 min would be billed, provided every minute started is charged (as is BTs policy). If rounding occurs at the end of every individual call then the following minutes will be billed:
   $2+1+4+3+1+9+6+1+2+7=36$
   As a result of this rounding policy, your bill will increase by 8.9%.

Consider the following: What would the result be if we were charged by the hour and rounding was done at the end of every individual call? For barely 32 min spent talking, 10 h would be billed! From this we can see that in order to make an informed decision on your tariff, information regarding the rounding policy of the company is necessary, particularly when you exceed your allocated free minutes of talk time.

We will now demonstrate how a real-life problem can evolve into a mathematical problem that requires an analysis of functions. Our goal here is to understand how the rounding policy of the company increases our bill.

Let us look at this example to get motivated:

> Twin sisters Maria and Anna have decided to make ten calls for exactly one minute to make the most of their tariff. Maria is very economical and savvy. She manages to hang up each of the ten times after exactly 59 seconds. Anna is not so efficient. She hangs up each of the 10 times after 1 minute and 1 second. What effect does this have on each of their bills?

Maria is charged for a total of 10 min and Anna for 20 min, even though she only talked 20 s longer than her sister. This seems unfair. To commence our analysis of this problem we will first represent the minutes Maria and Anna were charged using a chart. We have selected a bar chart with the x-axis showing the number of calls made by the twins.

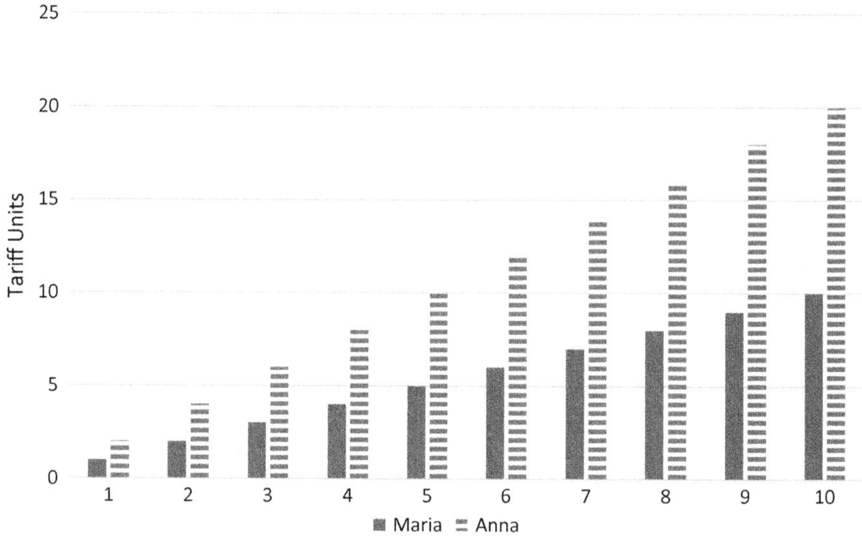

Fig. 1.2 Bar chart of total tariff units charged

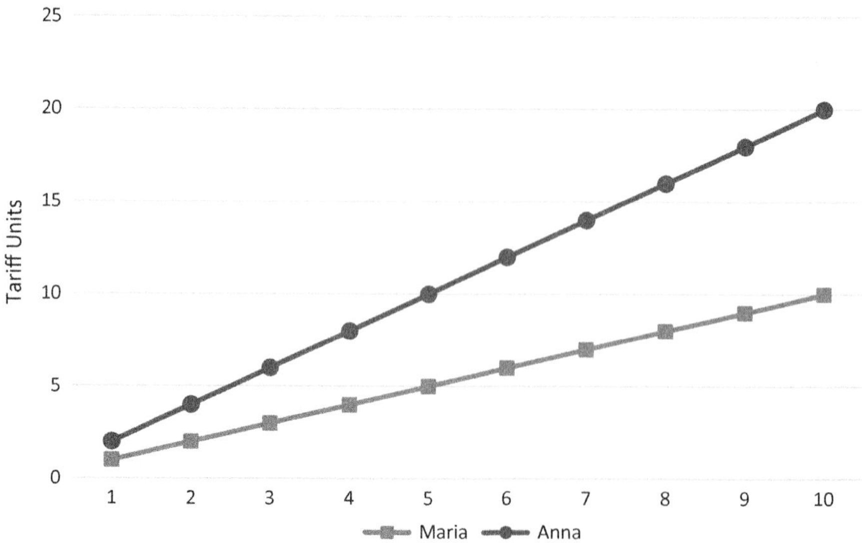

Fig. 1.3 Linear graph of total tariff units charged

Why is the following graph of a linear function not better suited to illustrate this?

This representation would make it rather easy to identify how much you would have to pay for 3.45678 calls instead of 3 or 4, as opposed to when the data was represented on a bar chart. However, how do you interpret the value of 0.45678 when referring to the number of calls?

Different groups of students or individual students working at home might have come up with different types of charts. This would be a good opportunity to consider the pros and cons of these charts and thus achieve a long-term learning outcome. Not everything a software program creates using a table of values makes sense! Or vice versa, it does make sense but it is first necessary to consider what information you need to glean from the chart before selecting a certain type of chart.

The above charts, outlining the total minutes charged, have made us, and hopefully also your students, more cautious. For example, in Fig. 1.2 we might be tempted to connect the top ends of the bars for each of the sisters to obtain two trend lines. This is another linear function, making it easy for us to compare charges. We did it this time (Fig. 1.3), but remain cautious.

For Maria and Anna we chose extreme values on purpose to attract attention. In the previous example (see Table 1.2) we used realistic values. Therefore, we shall revert back to these values and continue with this example using what we discovered to date. We will begin by creating a chart illustrating the cumulative minutes charged, assuming rounding takes place at the end of every call. Does your chart look similar to the one shown in Fig. 1.4?

On the x-axis the calls are consecutively numbered and on the y-axis the minutes charged are summed. How risky or helpful would it be at this stage to replace the top ends of the bars with points?

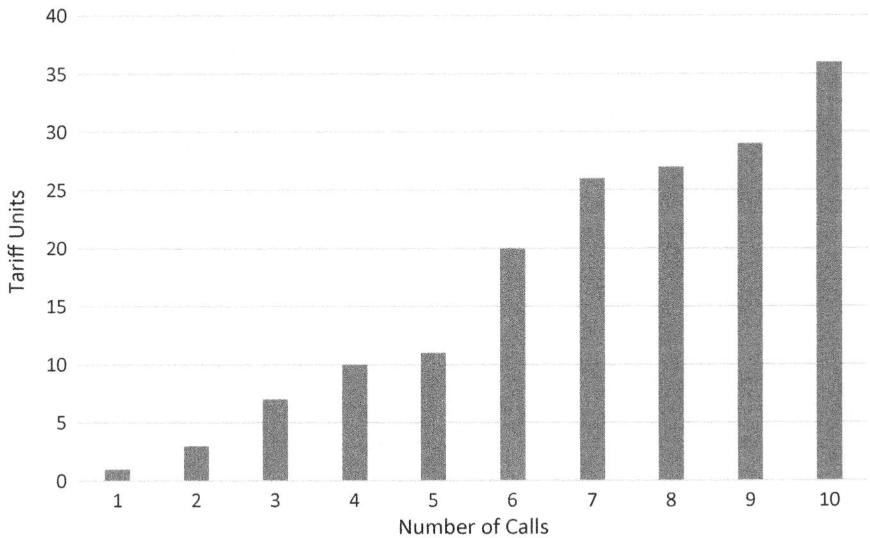

Fig. 1.4 Bar chart of cumulative tariff units charged

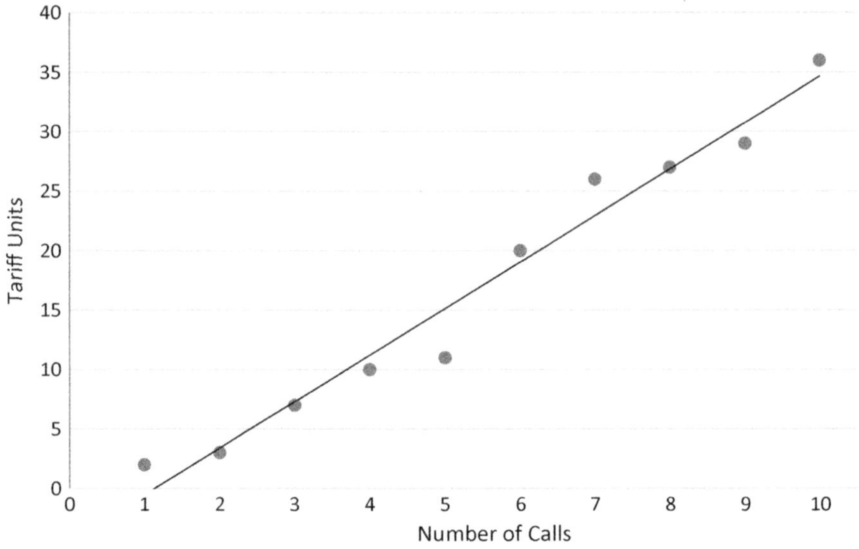

Fig. 1.5 Trend graph of cumulative tariff units charged

A trend line (a linear function) can be fitted to the points, as shown in Fig. 1.5, and we gain the following information using this trend line. For an average talk time of a little over 3 min (31 min 18 s ÷ 10), the trend line shows that approximately 3.5 tariff units per call have to be paid. Based on this information, once we know how many calls are made in a month, we can decide which plan is cheapest.

If we wanted to practice even more, we could use numbers generated by a random number generator for our talk time instead of the values given in Table 1.2 and then analyse it. After realising that the order in which the calls are made does not matter for our bill, we can sort the calls according to their duration. We will get a linear function as a result, which makes it easy for us to identify the cost and to give our recommendation.

The first improvement to our model, in this case taking into account the rounding off policy of the service provider, made us realise that rounding after each individual call resulted in a higher charge. The increase in cost, and consequently the bill total, is influenced by the talk time and the rounding off policy of the service provider. In summary, when rounding after every call, shorter calls which exceed a minute by only a few seconds cost more than longer calls, for which an additional minute will result in less additional cost units.

1.7 More Improved Model: Call and Data Charges

As a second step, we will now include the cost of data. At this stage, you could also include the cost of text messages or other additional costs. However, for now we will only consider how you can set up a mathematical model that includes calls and data charges.

It turns out that this model will also use a piecewise linear function. In this case however, it is a two-dimensional representation of a piecewise linear function using two variables (one to represent calls and one to represent data).

> Taking this into account, is the level of mathematics in lower secondary school extensive enough to conduct the search for a suitable provider in this case? What are your thoughts on this?

When comparing various plans, we once again note many different combinations. Again we have three different plans available to us from the BT advertisement:

Plan A—Flat rate 25 GBP/month with unlimited calls and data;
Plan B—10 GBP/month plan with 200 free minutes and 35 pence/min out of plan and 500 MB of free data and 10 pence/Megabyte (MB) out of plan;
Plan C—15 GBP/month plan with 500 free minutes and 35 pence/min out of plan and 2 Gigabytes (GB) of free data and 10 pence/MB out of plan.
Note: 1 GB = 1000 MB

Those who already know which services they are going to use can set out to look for a suitable plan. Those wanting to provide advice to others have to do a little more mathematics first. This way they get an overview of the plans which is easy to analyse. How do I go about that?

Below you will find, in algebraic and graphical form, each of these plans described as two-dimensional piecewise linear models (which can be represented using a three dimensional diagram—see Fig. 1.6).

function_1 (call costs, data) = 25 GBP/month

$$\mathrm{f}_1(x,\ y) = 25 \text{ with } x, y \geq 0$$

function_2 (call costs, data) = 10 GBP/month + 35 pence/min for all minutes in excess of 200 + 10 pence/MB for all MB used in excess of 500 MB.

$$f_2(x,\ y) = \begin{cases} 10 & x \le 200,\ y \le 500 \\ 10 + 0.35(x - 200) & x > 200,\ y \le 500 \\ 10 + 0.10(y - 500) & x \le 200,\ y > 500 \\ 10 + 0.35(x - 200) + 0.1(y - 500) & x > 200,\ y > 500 \end{cases}$$

$function_3$ (call costs, data) = 15 GBP/month + 35 pence/min for all minutes in excess of 500 + 10 pence/MB for all MB used in excess of 2 GB (2000 MB).

$$f_3(x,\ y) = \begin{cases} 15 & x \le 500,\ y \le 2000 \\ 15 + 0.35(x - 500) & x > 500,\ y \le 2000 \\ 15 + 0.10(y - 2000) & x \le 500,\ y > 2000 \\ 15 + 0.35(x - 500) + 0.1(y - 2000) & x > 500,\ y > 2000 \end{cases}$$

The following graph shows us when the talk time and data for Plan B and Plan C catch up with the flat rate plan (Plan A). In the graph, the talk time in minutes is represented on the x-axis, the data usage in MB is represented on the y-axis and the total charge in GBP is shown on the z-axis. By examining the graph, the relationship between Plan B and Plan C becomes apparent.

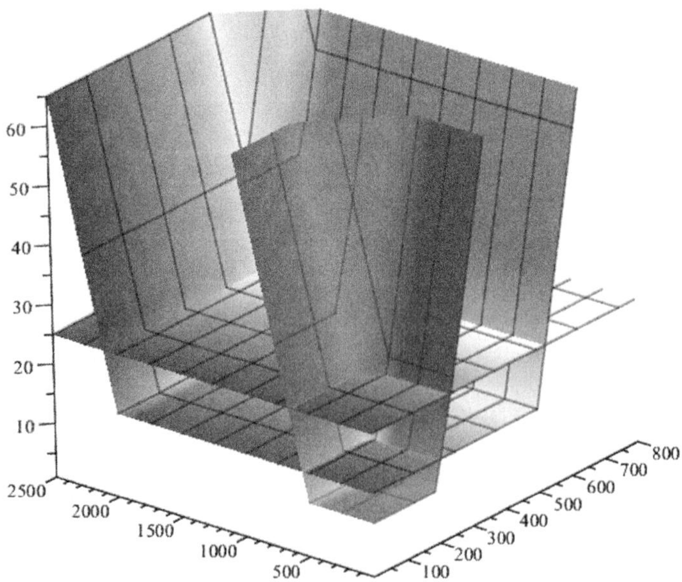

Fig. 1.6 Cost related to individual tariff plans when data is considered

Could students in lower secondary school correctly interpret this graph? The equations represent half-planes that intersect. However, this does not result in a (unique) intercept, but lines of intersection. This subject matter may be dealt with as part of analytical geometry in upper secondary school or at third level. It requires detailed background knowledge and cannot merely be taught using calculus or a graph. Nevertheless, an attempt at interpreting the graph can be made by students with a little help from their teacher. This way, students might also be exposed to some mathematical content that is to come in later years and thus be motivated to learn more.

At this point it becomes apparent that using a similar approach to that used for this initial extension of the model is not the most suitable approach for expanding the model further, for example if we wish to incorporate the cost of text messages or international calls. A four-dimensional display is not an appropriate tool here. As such, we have decided to break off at this stage, as we have already attained our pedagogical goals for this introductory chapter.

1.8 Mobile Phone Project: Reflection and Summary

At first, our goals for our mobile phone project were not very precise. However, we were able to illustrate with charts, and also in a verbal sense, how mathematics can help us make an informed decision regarding our choice of mobile phone plan, by granting us greater insights into the variety of information and offers available. We have only prepared the groundwork for how this could be done for data usage and texting. Continuing with this more complex problem would have gone beyond the mathematics required by lower secondary school students but we provided a glimpse of what would be involved.

On the other hand, using call charges as an example, it became apparent how attempting to achieve more accuracy or taking into account more details of the reality to be modelled can increase the complexity of the mathematical model. To stay on track, it is really helpful to recall the goals we set ourselves from time to time. Defining our goals also helps us decide where we need to set limits. If mathematics beyond these boundaries appears interesting, it is definitely possible to choose a question related to a real-life problem and deal with maths-intrinsic questions in a structured and general way.

Hopefully we have been able to explain typical aspects of modelling in mathematics in accordance with our goals for the introductory chapter of this textbook on modelling:

1. At first the goal needs to be set. This goal can be modified along the way, especially after available data has been analysed and initial calculations have been made. Sub-goals can also be defined. Clearly defined goals help us make decisions about the subsequent progress of our project.

2. Data and approaches for modelling one aspect of reality often have to be provided and sorted. In many cases the degree of accuracy to be obtained has to be determined in line with available resources. In the course of this book many of our examples will show that the degree of accuracy is often the determining factor for progress to be made in our project.
3. At the outset it often helps to reduce the level complexity and start with a smaller and more simplistic version of the entire problem. Of course the level of complexity of the real problem will remain unaffected by this. Therefore, it is advisable to increase the level of complexity later on until we reach an answer which we are convinced can withstand reality. There is no proof, in a mathematical sense, that the answer we found is good enough, only a check against reality will tell.
4. The end of the project is not determined by mathematics, or the solution obtained from the current iteration of the modelling cycle. It is a decision based on the subjective judgement of the quality of the result achieved. It is at the discretion of the people involved in the project to decide whether or not they have gained enough insight regarding their own goals or their questions about reality. When they are satisfied that this is indeed the case then they stop and implement the result.
5. Finally, let us analyse our findings:

   > Have you found the answers you were looking for in the course of this project?
   > Have you reconsidered your mobile phone contract and maybe even changed it?

   If you have, then you will be able to understand that this type of applicable mathematics education can, for example, result in changing actual consumer behaviour. Accordingly, this means that the expectations of the quality of our work are also higher than of mere textbook exercises. Whether a textbook problem is solved correctly or not may influence the student's insight and their grade. If a problem for a class project was solved correctly with sufficient accuracy, it could be life-changing, for better or for worse. Consequently, someone who obtains a solution to their project and goes about implementing it has to be prepared to take responsibility for it.

Some other aspects of applicable mathematics are not demonstrated very well by this example. We would therefore like to point out that the above is not an exhaustive list but we will discuss this further and provide more details at a later stage.

1.9 Concluding Remarks

We conclude this chapter with some observations about finding data and examples, and using the internet for this purpose. One of the most important advantages of teaching real world problems is the motivation generated in the students and teachers by the feeling that we are talking about things that are happening now and are relevant to us. So relevance and actuality are key considerations. Consequently, if you decide to put actual real world questions into the centre of a lesson, you will need actual real world data. Where do you, or better still your students, find this data? We feel it is not a good idea to search for this material solely in books or journals published years ago—nowadays looking for up to date real life data and situations means using the internet. This is why we use a lot of weblinks in this book. In addition, in order to make it easier for teachers and students to access the necessary data and information we have embedded, in the text, the full weblinks which will bring you to the exact location.

We know that all of the links given in our book will not be maintained indefinitely, or will be moved to other sites. This is a consequence of living with the internet—it is evolving and being constantly renewed. This is another meaning and consequence of actuality. We must learn to deal with this eventuality. We hope to lower the risk of going to dead links by giving additional information about the group that offers the information. For example, we write that the ISTRON group has published basic ideas about teaching real world problems at: http://istron.uni-koblenz.de/istron/. This website was located at the University of Koblenz for many years but is no longer active because one of their group members, who maintained the site, has moved to another university. However, we think if you type something like "ISTRON mathematics" into a search window you will find what you need. In this way, we acknowledge the work of the group and show you how to find useful information for yourself and your students.

References

Blum, W., Galbraith, P. L., Henn, H.-W., & Niss, M. (Eds.). (2007). *Modelling and applications in mathematics education—The 14th ICMI study*. New York: Springer.

Cheng, L. P., & Chua, K. G. (2015). Learning through "MOBILE PHONE PLAN". In N. H. Lee & N. K. E. Dawn (Eds.), *Mathematical modelling: From theory to practice* (pp. 183–192). Singapore: National Institute of Education.

Davis, P., Hersh, R., & Marchisotto, E. A. (2012). *The mathematical experience* (study ed.). New York: Springer.

Schoenfeld, A. H. (1992). Learning to think mathematically: Problem solving, metacognition, and sense making in mathematics. In D. Grouws (Ed.), *Handbook of research on mathematics teaching and learning* (pp. 334–370). New York: McMillan.

Stillman, G. A., Kaiser, G., Blum, W., & Brown, J. P. (Eds.). (2013). *Teaching mathematical modelling: Connecting to research and practice*. Berlin: Springer.

Chapter 2
Motivation—Why Teach Applicable Mathematics?

Teachers often ask the following questions: How will my students benefit from applicable mathematics? What is the added value for my students when they engage in applicable mathematics? How will engaging in applicable mathematics benefit me as a teacher? How much additional work will it mean for me as a teacher?

We have been discussing the issue of motivation with mathematics teachers and mathematics pedagogues for many years, trying to find conclusive answers to these questions. We firmly believe that we have found some answers and we will summarise them for you in this chapter. Further, we believe the insights that we offer have a universal appeal. However, from our experience and many discussions with mathematics teachers, mathematicians and mathematics educators, we know that what is convincing to us is not necessarily so to others. Experience suggests the strongest arguments do not always result in a change of behaviour, even when they are accepted by teachers. Those who already use applied problems and approaches in mathematics education, or want to try them out or want to learn about them during their studies, do not need to be persuaded. What they need is a little encouragement. On the other hand, those mathematics teachers who have been teaching in a conventional way for years and do not wish to engage with applicable mathematics education will probably not be persuaded by our arguments.

Questions such as those listed above to frame this chapter need to be dealt with in a systematic way. As regards supports for the implementation of applicable mathematics education, mathematics teachers in German-speaking countries have been very well served for several decades by two networks: MUED[1] (www.mued.de) and ISTRON[2] (http://istron.uni-koblenz.de/istron/). In these countries such ques-

[1]MUED is an organisation of mathematics teachers, from German-speaking countries, devoted to the improvement of mathematics teaching and learning thorough the development of applicable classroom-ready mathematics lessons.

[2]The main aim of the ISTRON group is to contribe to the improvement of mathematics education through the development of reality-based classroom-ready materials for the teaching and learning of mathematics.

J. Maaß et al., *Mathematical Modelling for Teachers*, Springer Texts in Education, https://doi.org/10.1007/978-3-030-00431-6_2

tions are addressed during MUED-conferences and other continuing education events for teachers. Similarly, many mathematics teachers get an opportunity to pose such questions regarding applicable mathematics education at continuing professional development (CPD) events at ISTRON conferences and similar events.

The situation in English-language countries is less organised and coherent when it comes to dedicated supports for reality-based mathematics education, but they do exist and can be accessed by teachers. Generally, one needs to look at this country by country as there are few, if any, dedicated pan-national networks or associations. However, many national mathematical organisations and associations provide reality-based materials for use in schools and are accessible by teachers everywhere through websites and published materials. Here we also recognise the role of international conferences such as the Congress of European Research in Mathematics Education (CERME) and the International Congress on Mathematical Education (ICME), that host themed sessions and topic study groups on problem solving, modelling and applications of mathematics. In this regard the International Community of Teachers of Mathematical Modelling and Applications (ICTMA) merits special mention. Since 1985 there has been a community of mathematics educators dedicated to the promotion of applications and modelling at all levels of education including schools. ICTMA organises a biennial international conference and is responsible for a constant output of high quality material and publications on modelling in mathematics including school mathematics (http://www.ictma15.edu.au/). We will direct mathematics teachers to other appropriate sources and resources as they progress through the book.

Once again we ask for your active cooperation here. Please record your own point of view on this discussion;
Have you previously conducted applicable mathematics lessons?
Was it a good or a bad experience for you?
Are you for or against it?
Please write down your most important arguments for and against applicable mathematics lessons and return to these notes after you finish reading the following pages.
Do you still have the same view?
Or more importantly, has your point of view changed after reading on?

2.1 Long Term Positive Image of Mathematics

Let us begin by considering the typical student perspective on mathematics lessons. Students often ask the question "*Why should we study this material?*" Typical answers given by teachers such as, '*it is on the curriculum*' or '*you will understand later when you have studied more mathematics*', are not very well received by students because they lack persuasiveness. We are perfectly aware that it is not always feasible to provide persuasive arguments that would be acceptable to each and every student regarding what should be studied for every subject. It would be too time-consuming in the first place. Nevertheless, it makes one think that something must be wrong if most students feel this can never be done during their time in mathematics education. It is even more regrettable when most adults consider that their school mathematics education was ineffective or useless. Frequently, international empirical studies on how adults see mathematics have produced negative findings. For example, a number of studies have shown that conventional mathematics education mainly produces negative effects such as poor knowledge and negative feelings towards mathematics. Generally, many people have bad memories of school mathematics. For more details on this see FitzSimons et al. (1997).

Does it matter to you, as a teacher, what people think of mathematics? Can mathematics teachers afford to ignore the poor image of mathematics that is known to exist among the wider population? Such issues impact on mathematics teachers' classroom practice and must be a matter of concern. We know the majority of mathematics teachers dedicate a lot of their expert knowledge, time and energy to their classes. Despite this commitment outcomes are not always positive. Empirical studies focussing on the mathematical skills of learners, particularly adults, paint a bleak picture. They show, on average, that far too little of the mathematical content studied at school produces a long-term learning effect. We firmly believe that more mathematical knowledge and skills can be taught and retained, on a long-term basis, by implementing small changes in mathematics classrooms. Applicable, reality-based mathematics education is an important step in that direction.

One way to achieve this end is to implement a reality-based project once or twice a term/semester. We would like to stress that we are not saying all classes should be conducted in a hands-on fashion. We acknowledge that this is only one of the many facets of mathematics that students should become accustomed to during their mathematics education.

2.2 Understanding Mathematics Better by Knowing More Mathematics

Another important question that often arises is: What is the point in engaging with several branches of mathematics in mathematics education? It is important for us to appreciate that mathematics is a very versatile science with many different branches and topics that have connections with other sciences, professions and everyday life. It has a long history and a very specific formal-axiomatic methodology in contrast to other sciences. For example, an algebraic expression on groups or shapes is neither put to an empirical test, as we would do in the social sciences, nor to an experimental test, as we would do in the natural sciences.

As we, and many of our colleagues in mathematics and mathematics education departments, see it, the main aim of mathematics education is for students to familiarise themselves with, and appreciate mathematics for its versatility and uniqueness. Furthermore this viewpoint is shared by many people responsible for drawing up mathematics curricula worldwide such as Ministries of Education. This way there is a greater chance that students will encounter one or more aspects of mathematics which will give them personal and positive access to mathematics. To do this effectively, we need to expand their mathematical horizon and their understanding. Mathematical understanding is generative, allowing students to integrate their knowledge and develop problem solving skills (Blum et al. 2007; Schoenfeld 1992). On the other hand, performing only mathematical operations (calculating something or in more general terms following a predefined path to solve certain types of problems) restricts the image of mathematics and its accessibility. When the focus of mathematics education is predominantly, or exclusively based, on performing mathematical operations, mathematics becomes harder to grasp and less accessible to students. Ultimately, this type of instruction proves ineffective, as evidenced by the poor disposition and performance of many of those who experienced it (Blum et al. 2007; Schoenfeld 1992).

Whenever we advocate applicable mathematics education and greater knowledge of mathematics, we are asked to consider what should be omitted instead. After all, the number of hours of mathematics instruction remains constant at best or even decreases sometimes. The contrary argument always puts the question: When something new is added, does something else not have to be omitted? The general feeling is that something should be omitted but this always leads to a difficult discussion.

The key idea in this discussion is the breadth of subject matter to be studied. It is no longer possible for anyone, even the most talented mathematicians, to be familiar with all the new propositions and theorems, from all the various branches of mathematics, let alone understand them all. This increase in mathematical knowledge over a short period of time is almost irrelevant for mathematics in schools. New mathematical insights are often impossible for students to grasp, even in a simplified form. Unlike the natural sciences and humanities, scientific advances do not lead to new chapters in textbooks or topics in mathematics curricula.

So why then do mathematics teachers still feel more overwhelmed about the breadth of their subject matter compared to teachers of other subjects? We believe the reason many mathematics teachers feel overwhelmed by this breadth of material lies in their mathematics-related training. Mathematicians are used to answering questions completely, getting to the bottom of associations and connections and considering a systematic order of their subject matter as particularly important. In other words, good mathematicians do not like it when an explanation is rudimentary and incomplete. For example, when the curriculum contains a list of topics to be covered (such as being able to manipulate the subject of a formula or being able to solve 2 equations with 2 unknowns), most mathematics teachers interpret this as a mathematician would, by exploring as many types of formula manipulations or solution procedures as possible. They have their students practice these problems excessively, so that they can solve similar problems during examinations. This is how a concise curriculum requirement impacts classroom practice, by turning a simple topic into an extensive classroom activity which manifests itself in some school systems as a distortion of good practice, which is not the intention.

We want mathematics education to change fundamentally. So far conventional mathematics instruction has not produced the desired effects, despite all the effort put in by mathematics teachers. The average levels of mathematical knowledge and skills, among students, are not as good as desired nor do they endure long-term. Furthermore, as a result of conventional mathematics education the image of mathematics among the general public has suffered—this message is relayed time and again through test results like PISA, or different National Standard tests. This state of affairs is no secret and it has prompted many efforts to improve mathematics education worldwide.

We do *not* propose adding another requirement, that of applicable mathematics education, to the numerous existing requirements. Instead, we suggest returning to what is really important from a pedagogical point of view. It is easy to learn what you enjoy doing! We therefore suggest focusing on motivation. There is no point in teaching more of the same with even more fervour. Just try a different approach! Having gone through university-level mathematical studies, the sequence of definition—proposition—proof—application has become second nature to us (and most other mathematics teachers). This approach to teaching mathematics was advocated and widespread in our student days. We now claim that this sequence is discouraging and repressive to most students in schools (and also universities—but this is not our topic here). This is particularly true if the sessions on applicable mathematics are limited to practicing certain types of problems. In such cases the teacher only gives quick explanations on the other aspects of the problems or jots something down on the blackboard (often accompanied by an apologetic comment that the proof on the blackboard is not part of the next test anyway).

We therefore propose not to start a mathematics lesson (a new topic to be precise) with a definition, but with a motivating phenomenon or an interesting question taken from everyday life. In the course of answering questions that arise along the way you should introduce or develop the mathematical tools needed for this purpose. When a modelling session requires some aspect of mathematics which has not been explained, proven or practiced yet, this is not the end of the project. On the contrary, it is a wonderful opportunity to discover or develop a previously unknown aspect of mathematics in a motivating situation. During the long history of mathematics a new and useful theory has often evolved from a practical problem in physics, economics or pure mathematics. For example, Kepler's computation of planetary orbits is one of many motivational examples generated by an interesting question, which in this case led to long and laborious computations.

2.3 The Mathematics Teacher's Perspective

At this point we return to the chapter framing questions: How will engaging in applicable mathematics benefit me as a teacher? How much additional work will it mean for me as a teacher? By posing such questions and engaging with them, we reveal something about how we view the role of mathematics teachers. Teachers, students, schools and society in general benefit greatly if mathematics teachers act as the responsible and mature citizens upon whom students can model their own behaviour. If they act as an example in this sense, they can make a substantial contribution as pedagogues.

> Once more we ask for your active cooperation. Please note the pros and cons you expect (or already experienced) when conducting applicable mathematics lessons.
> If nothing comes to mind right away, here are some key ideas to work with: the situation in class, the joy you feel when teaching and preparing for class, the level of student engagement and preparation time.

It is common knowledge that teachers enjoy teaching in a hands-on way and writing about it. This enjoyment arises because, on the one hand it is much more enjoyable to teach students who actually want to learn and are thus motivated to follow the lesson, while on the other hand this approach gives us the opportunity to learn more about the world we live in ourselves. In short we are interested in how our world works and applied mathematics, modelling and simulation, helps provide us with some answers. Documenting our quest and looking at it from the perspective of a mathematics teacher educator opens up a path for teachers and students to look for their own answers and arrive at their own conclusions.

For example, thinking about ways to save energy and protect the environment has resulted in many related proposals, which at least made us, and probably also our students, better understand the problems at hand (but the focus of this book is on how we can benefit as teachers). Even a nuisance like being stuck in traffic at the entrance to a tunnel can be turned into a project for applicable mathematics education. Developing it as an activity for an applicable mathematics lesson has the potential to make the wait time more endurable for an active mathematics teacher who can engage in real time if s/he is inconvenienced in this way.

The following are some useful websites that may be helpful when teachers are trying to incorporate applicable mathematics in their classroom:

Math Models—http://www.mathmodels.org/
Stepping Stones Projects—http://www.indiana.edu/~hmathmod/projects.html
Plus—https://plus.maths.org/content/teacher-package-mathematical-modelling

Furthermore, you can look up the MUED and ISTRON websites for a list of topics with proposed lessons that have been put to the test and published in the German language.

The additional pleasure derived from research and teaching comes at a price—it requires a lot of additional work. To us it is a part of our work that we enjoy doing. Nevertheless, we are aware that there are some teachers out there for whom the additional time invested in preparing their lessons poses a problem. New teachers, who start their teaching career thinking that every lesson should be prepared thoroughly, will soon reach their limits. When this happens we suggest new teachers prepare a few selected lessons a month very thoroughly so that they can teach good applicable mathematics lessons systematically. This investment in preparation time will certainly pay off. You can balance the extra investment in time spent preparing applicable mathematics lessons by teaching some classes within your comfort zone, which does not require as much preparation time. Often the additional effort poses a problem for experienced teachers. Perhaps they may have cut down on preparation time because lesson preparation has become routine. Even though lots of useful proposed reality-based lessons already exist, it still requires a little time and effort to gather some information on the topic and adapt the proposed lesson for one's own class.

We do not wish to underplay this additional effort because it is substantial. However, we do suggest that these teachers should be willing to do something positive so that they themselves and their students are able to spend at least a few classroom hours actually enjoying their mathematics lessons. Maybe this can be compared to hiking in the mountains. Hiking no doubt requires more effort than relaxing and sunbathing in your deckchair, but such effort will be rewarded with magnificent views and a sense of achievement. Remember, an important aim of education is training someone to take responsibility for his or her actions. According to numerous published sources on goals, curricula and also statements on teaching methods, using their mathematics to bridge the gap to reality is a step in the right direction.

References

Blum, W., Galbraith, P. L., Henn, H.-W., & Niss, M. (Eds.). (2007). *Modelling and applications in mathematics education—The 14th ICMI study*. New York: Springer.

FitzSimons, G., Jungwirth, H., Maaß, J., & Schlöglmann, W. (1997). Adults and mathematics (adult numeracy). In A. Bishop, et al. (Eds.), *International handbook of mathematics education* (pp. 755–784). Dordrecht: Kluwer-Verlag.

Schoenfeld, A. H. (1992). Learning to think mathematically: Problem solving, metacognition, and sense making in mathematics. In D. Grouws (Ed.), *Handbook of research on mathematics teaching and learning* (pp. 334–370). New York: McMillan.

Chapter 3
Adapting Textbook Problems to Create a More Reality-Based Mathematics Education

How can typical mathematics textbook problems be turned into motivating reality-based problems? In this chapter you will learn how textbook problems can be slightly altered to make them more open-ended and we will discuss how to solve these. Problems on various topics and for various educational levels are also discussed.

Historically mathematics textbooks have been a key resource for mathematics teachers. However, textbooks and other curricular materials must be aligned purposefully with the curriculum goals but this is not always the case when we compare traditional mathematics textbooks with the objectives of reality based mathematics education. There are some curricular materials that meet this requirement of reality-based mathematics, such as Bowland's maths project (http://www.bowlandmaths.org.uk/index.html) and the Mathematics in Context (MiC) project (http://mathincontext.eb.com). However, many problems presented in textbooks are not purposefully aligned with applicable mathematics education but some of these problems can be adapted for use in the applicable mathematics classroom. In this chapter we offer some suggestions and guidance for adapting traditional problems that will enable you to implement this new approach.

As a mathematics teacher you regularly select, adapt and assign problems for your students. At first glance this seems an easy task as long as the problem is easy to create or adapt. For example:

> Calculate $5+8=?$
> or
> What real values of x satisfy the following equation?
> $x^2 = (x-2)+(x+2)$

In the first case, such a simple addition problem can, at worst, turn into a problem if a typographical error occurs, for example instead of $5+8$ it says $5-8$, and negative numbers have not yet been introduced. Obviously, if you use 'spending money' as the context and talked about 'getting into debt' then the meaning would become easier to understand in the case of the subtraction problem.

J. Maaß et al., *Mathematical Modelling for Teachers*, Springer Texts in Education, https://doi.org/10.1007/978-3-030-00431-6_3

In the second problem there would not be any solution if the typographical error involved using multiplication (×) instead of addition (+) between the brackets. In this case the problem would become $x^2 = (x - 2) \times (x + 2)$ and therefore makes no sense, as zero cannot equal −4. The consequences of such a typographical error depend on the classroom situation. If your intention is to remind your students during practice for tests or exams that quadratic (and other) equations sometimes do not produce a valid result, then the task fulfils its purpose. However, if the typographical error occurred during the very first problem you presented on quadratic equations and multiplication (×) was typed instead of addition (+), the result of what should be a simple calculation of x for the reduced equation $x^2 = 2x$ (therefore $x = 2$ or $x = 0$) is perhaps one that has not been mentioned before—the case of no solution. This result can be discussed right away or at a later stage, but it will probably disrupt your planned lesson.

3.1 Challenges in Creating Mathematics Problems

The two previous examples should remind you that creating mathematics problems is a challenging task, especially if you intend to assess certain facts, knowledge or skills that you decided upon in advance. Even if problems using only mathematical symbols are concerned, which are apparently easy to create, it is not always easy to present what needs to be done in a simple, precise and accurate fashion. After all, the type of problem often determines the way it is assessed afterwards. If the problem was ambiguous in the first place, it is harder to determine if it was solved correctly or not. This makes correcting and marking it that much harder.

The situation gets even more challenging when problems are written mainly as text. Maybe you had a similar experience to us when you wrote your first word problem. In our case, we believed the problem was written accurately, was easy to understand, and motivating as it related to people's everyday lives. Nevertheless, some of our students had a different view on this. They did not interpret the problem in the way we had intended; did not consider it more motivating than the ones we tried to set it apart from; and some even found the problem confusing or impossible to solve.

Despite all the challenges previously mentioned, we still aim to encourage and enable you to use new types of problems suited for applicable mathematics education. Before addressing this challenge, we would like to draw your attention to something we learned during a training session for mathematics teachers. In this session, the text used in a selection of textbook word problems seemed odd to the group of mathematics teachers. Hence they asked us to consider how the language used could be improved. To do this we split up into teams and tried to rewrite typical word problems taken from mathematics textbooks so that the language was improved from the viewpoint of the teachers but still remained easy for the students to understand and solve. The teachers then tried to solve the problems created by the other groups, often in vain it turned out, as they could not extract from the eloquently rephrased text what they were supposed to do.

From this experience it was concluded that elegant prose text does not always work. Consequently, we would like to stress that textbook problems often use a sort of artificial language which does not sound too elegant, but combines precision, unambiguousness and brevity in an optimum way. Conversely, can you imagine a poem or prose that fulfils the three criteria of precision, unambiguousness and brevity to a high degree? To be specific, take a poem, a travelogue or a novel and try to rephrase it in such a way so that it is as precise, unambiguous, and brief as possible. We do not intend to criticise your literary talents here, but we assume that you will not win any literary prizes for the text you rephrased for mathematical purposes. The reason for our assumption is simple: the aims and assessment criteria for suitable textbook problems and beautiful literary texts are very different.

Summing up our initial considerations on how to create textbook problems, one could say it is not easy to do it well. The wording requires careful attention as even small mistakes, for example positive/negative signs, can create big problems. The high degree of precision, unambiguousness and brevity required for well-written textbook problems is a determining factor in deciding whether or not the problems are fit for purpose. Despite criticism from outside the profession and from other disciplines, it is important to maintain this kind of specific language for these reasons.

3.2 On the Didactical Merits of Textbook Problems: New Perspectives

Up to this point we have avoided writing about the didactical merits of textbook problems on purpose. Apparently, there are well-written problems that are lacking in didactical merit, and conversely, there are some that have the best didactical intentions but are not well written. In other contexts the relationship between form and content is also well worth considering. Certainly one cannot be independent of the other, in fact, they are interdependent.

Once again we ask for your active cooperation here.
What exactly is the didactical merit of textbook problems?
Write down what didactical merits a problem used in your mathematics class can have?

We have identified two important didactical advantages. Textbook problems provide opportunities for practicing and also understanding worthwhile mathematics.

Practice: Frequently, students are presented with textbook problems so that they can practice what they have studied. For example, once a formula for calculating interest is covered in class, a series of problems which requires students to calculate interest rates ensues. Likewise, if the teacher explained how two linear equations in two variables are to be solved, this is often followed by a series of problems on the topic.

Understanding: The aim of many of these textbook problems is that in the course of solving them, students understand what had 'merely' been explained to them before. For example, if the Pythagorean theorem and its proof is on the board but students do not understand its real meaning just yet, some calculations can often do the trick.

In practice, some textbook problems are better suited than others for each didactical purpose. Some are too simple, others too complex, some are not well written, and others are typical of the type of problem that is to be practiced. Other textbook problems can be well written in all regards, have high didactical merit and yet students cannot solve them during the important class or school test, even though they did well on similar problems during practice and revision sessions.

There is also a third criterion that we recommend when evaluating problems: their success when put to use. This success may be experienced by students and/or teachers during class—or empirical research can be done on it. The first case is about a specific success right there and then. Does the problem provide the appropriate opportunity for practice or contribute to student understanding? In the second case, general statements can be made about how easy the wording of the problem is to understand or about the correlation between classroom instruction, educational objectives, the mastery of certain skills and the way in which a problem is presented.

Textbook problems considered useful from a didactical point of view are usually valued as textbook problems for long-term learning. This involves a change of perspective when evaluating a problem. A problem may seem very different when viewed from the perspective of someone who created it as opposed to from the perspective of someone who is supposed to solve it. If we make long-term learning our top priority in judging the didactical merit of a problem, it becomes necessary to put these problems to the test and do some empirical research on them. Only afterwards can we make our final assessment. Here, a longitudinal survey would be our best option to look into the 'sustainability' factor, identifying which problems contributed to supporting long-term learning. In the course of the survey, we would need to question the students over a longer period of time (or quite some time after the classes in question took place) and test or interview them again to find out which knowledge, skills or competencies they acquired permanently.

We shall not go into further detail on conventional textbook problems at this stage. Should you wish for more details on this, please read some of the publications listed here (Pehkonen et al. 2013: https://core.ac.uk/download/pdf/33981175.pdf; De Corte et al. 2000: http://math.unipa.it/~grim/Jdecorte.PDF).

In the selected readings referred to above you will find a discussion of alternative approaches to problem solving. In this context you will have noted keywords such as 'open problems'. If this is the case, then you will be particularly interested in our suggestion that opening up mathematics problems may lead to more hands-on mathematics education. Open problems differ from traditional or closed mathematics problems because they allow for multiple correct answers through different routes, and engage different skills, such as practical and process skills. How does this approach work? Here are some examples.

> Rakesh wants to buy a book that is €14.90. He has €20. How much money does he get to keep?

Students who work this out correctly will then write: Rakesh can keep €5.10, and they might even underline it twice for emphasis. We suggest rephrasing the problem as follows:

> Rakesh received the book 'Harry Potter and the Philosopher's Stone' by J. K. Rowling as a birthday present and he enjoyed reading it. He has now come across another book by J. K. Rowling on the internet; 'Harry Potter and The Chamber of Secrets'. On Amazon, the book is €14.90. Rakesh still has €20 his uncle gave him for his birthday. What should Rakesh do? Discuss in small groups what advice you would give Rakesh.

Using this kind of wording, students' advice to Rakesh is more open. The calculation that he gets to keep €5.10 if he buys the book is only a minor matter, one solution among many. Other solutions could be that Rakesh should borrow the book from a friend or the library. He also ought to check if a used book is available somewhere. Another option would be to put it on his wish list for Christmas. Certainly more ideas and options can still be found.

It is obvious now that the calculation no longer takes centre stage. The question of how much money Rakesh gets to keep, if he buys the book, is one of many possible questions that may arise. Possibly the result of the calculation can be motivation to look for cheaper solutions. This is quite realistic, but so far it is not customary to make time for such considerations during mathematics classes. The time spent looking for other ways to get to read the book is not conventional mathematics education but time spent learning life skills. Time spent in this fashion reduces the time available for practice problems on the same topic.

This is the key issue, in general, in the didactics of mathematics that we would like to discuss with you: What is really important in mathematics education? If students realised how they could make good use of the advice they gave to Rakesh for their own lives, we believe this approach is generally in line with national educational goals for mathematics education. To be more precise, it would be beneficial if students learned how they can assess the quality of their own advice and remember how they can obtain a book they would like to read. At this point we would like to stress once again that we do not suggest teaching in this hands-on, open-ended fashion all the time.

Let us look at another example. Problems on calculating interest often involve questions asking us to calculate how much money somebody who invests €100,000 in a bank for 10 years at a fixed interest rate will have at the end of that period. Often the technical term 'fixed-term savings account' is avoided, and it is worth nothing that the actual interest rate nowadays tends to be lower than in previous textbook problems. We instead suggest rewording the problem as follows:

> Sylvia receives €5,000 from an aunt for her 14th birthday. She is supposed to place the funds in a fixed-term savings account for 5 years, so that she will have as much money as possible on her 19th birthday. What advice would you give Sylvia about investing her money?

If students require more scaffolding, then the following suggestions could be provided. Look up various banks and do some research on the internet (online banks). Discuss the offers you found in small groups and make a proposal. Your proposal has to show how much money Sylvia will have after 5 years and which risks and uncertainties the planned investment will involve. Please insert the interest rate, fees and pay-out (future value of the investment) offered by the bank into a table and check if your figures are correct (this can easily be done on a spreadsheet). After discussing the student presentations, one proposal is to be chosen as the best option. The group whose proposal has been chosen gets a prize!

In this way we can turn a standard textbook problem into a real-life project. We shall not repeat again the pros and cons regarding time, the issue of extensive material to be covered and general educational objectives. However, we feel that it would definitely be a good thing if schools helped students learn how to manage their money better. This real-life project also involves making interest calculations as well as understanding the small print containing the terms and conditions of deposits and loans.

3.3 Improving Reading Skills in Mathematics Education

Once again we are confronted with the following question regarding reading skills: Does this question belong in mathematics education? Should this skill not be acquired in language classes and then used in mathematics classes if needs be?

In our view, one single reading skill for all kinds of texts does not exist. Different types of texts require specific reading skills. We interpret the situation as follows: specific background knowledge is essential to understanding subject-specific texts. People who can read texts on tax law, the synthesis of certain organic compounds in petroleum, the diagnosis of liver diseases or alternative proofs of the four colour problem, without foreign words and technical terms troubling them, often cannot really comprehend them without specific background knowledge. For example, a text on training schedules for professional cyclists requires different background knowledge than a text on the literary quality of a new play.

We therefore submit that no school subject, other than mathematics, makes provision for the teaching of mathematics-related reading skills. Of course classes in the mother tongue (English, French, Spanish, German) and to some extent the natural sciences and technical subjects, contribute their share to mathematics-specific reading skills but it is essentially in mathematics classes that this skill needs to be acquired. For us, it also goes without saying that this sort of skill can only be acquired when actually dealing with texts with mathematical content in mathematics classes. For example, texts on mobile phone plans or ads for investments or bank loans contain such mathematics-related information.

Teachers might ask, is this not a bit too much for our unprepared students? Let us first deal with this typical objection on the teacher's part before moving on to our next example. The simple answer is yes! Someone who gets to see a loan contract, an insurance contract or a contract with electricity suppliers or phone companies for the first time, will most likely feel overwhelmed by it. But the consequences of not dealing with such matters in school are not very pleasant. Those who are not trained to handle money in real-life situations outside of school and who do not learn how to manage their money wisely, need to trust in consulting services or ads or learn mathematics on their own later in life. In our view the proper thing to do is to start at an early stage, and in small steps, to incorporate reality into mathematics education using texts with mathematical content. This approach provides an opportunity to learn and practice together how to extract the information needed and analyse it according to mathematical principles, so as to provide the basis for our decisions.

3.4 Learning to Solve Problems Independently Using Altered Textbook Problems

> To begin this section, we will ask for your active cooperation.
> Choose a typical textbook problem and add some information that has nothing to do with it. Use your own school textbooks for this exercise.

Meanwhile let us use a virtual problem for illustration purposes:

> A shipping company transports parcels from Vienna to Linz (180 km) and on to Salzburg (another 120 km to the west). What distance does the delivery truck have to cover between Vienna and Salzburg?

We have added some additional text unrelated to the problem and this is shown below:

> A shipping company transports 264 parcels from Vienna to Linz (180 km) and 125 parcels on to Salzburg (another 120 km to the west). What distance does the delivery truck need to cover between Vienna and Salzburg?

Now examine this altered version. How many students add parcels instead of kilometres? How many add parcels and kilometres? You might be familiar with a version of the following problem:

> The journey of a 53 year old bus driver starts at the central station. 32 passengers board the bus there. At the first stop 3 passengers get off and 5 get on. At the second stop 7 passengers get off. How many passengers are on the bus before the third stop?

Were your students confused by the unnecessary information regarding the bus driver's age? Do they add 53 (the age) to the total number of passengers? How would your students react if the final question was a different one, for example:

> The journey of a 53 year old bus driver starts at the central station. 32 passengers board the bus there. At the first stop 3 passengers get off and 5 get on. At the second stop 7 passengers get off. How old is the bus driver?

For analysis purposes we suggest another version of the problem:

> The journey of a bus driver starts at the central station. 32 passengers board the bus there. At the first stop 3 passengers get off and 5 get on. At the second stop 7 passengers get off. How old is the bus driver?

Do your students calculate the age using the number of passengers? There are students who would actually do this. Here are the results of an empirical study. In 1989 the book "How Old is the Captain" by Stella Baruk caused a stir. This book reports on an experiment involving second and third-graders (7–8 year olds) who were asked to solve the following problem:

> On a boat, there are 26 sheep and 10 goats. How old is the captain?
>
> (http://www.seuil.com/ouvrage/l-age-du-capitaine-de-l-erreur-en-mathematiques-stella-baruk/9782020183017)

76 out of 97 students answered '*36 years*'. A wider net was cast for the original French research including more problems and students of varying grades and the result was similar. Other studies in the USA and Germany paint a similar picture. While only 10% of pre-school students attempted to solve such 'captain' problems, a higher percentage of early elementary school students (6–9 year olds) did so. From grade 5 onwards, the percentage attempting such problems once again sinks back below 50%.

(https://www.yumpu.com/de/document/view/10048938/53-kapitansaufgabe).

There is a third quick and easy way to alter textbook problems and create more open mathematics education so as to make classes much more interesting than usual. Simply replace the information by a question. This way the problem turns into something new and open-ended:

Example 1
Standard: Assume a right angle triangle with $a = 4$ cm, $b = 3$ cm. How long is the other side c?
New: Assume a right angle triangle with $a = 4$ cm, $b = 3$ cm. How long is the other side c? For which angle γ can we solve this problem using the Pythagorean theorem? (α is the angle opposite the side a, β is the angle opposite the side b and γ is the angle opposite the side c).

Example 2
Standard: There is a straight line $f(x) = 4x + 5$. Where does it intercept the x-axis and the y-axis?
New: There is a straight line $f(x) = a * x + 5$, where a is a real number. What values can a take so that the line intercepts the x-axis but not the y-axis?

Example 3
Standard: Assume a linear real function with values $(-3, -3)$ and $(2, 2)$. Show that at least one point on the x-axis is also part of the function.

New: Assume a linear real function with values $(-3, -3)$ and $(2, 2)$. Under what circumstances does it intercept the x-axis at exactly one point?

Initially we have chosen examples unrelated to reality to demonstrate that small alterations to conventional problems can lead to new perspectives, thus making mathematics class more interesting. In some situations it makes sense to give such problems as additional practice or homework.

3.5 More Examples for Different Topics and for Different Grades

Let us assume we were able to convince you of the merits of this new approach and you would now like to start teaching in a more active and open-ended way. Please consider for which grade(s)/age group(s) you would like to try this approach and for what topic. At this stage we suggest you have a look at internet resources and journals about mathematics education (for concrete hints see Chap. 7).

We will give you some more suggestions now on how you can turn textbook problems into small classroom units. To this end, we will select various problems from randomly chosen textbooks for various school types, including elementary, middle and high school, and grades. How can we adapt these problems?

> A sports club rents a bus for a tour for €120. The cost is to be shared equally between the passengers. If two additional people come along, the cost for each individual would have decreased by 25 cent. Determine the number of people who participate.
>
> (Dorner, Grade 5, Problem 11c, p. 107)

The usual approach for solving such problems would involve translating the text into equations to determine that initially 30 participants owed €4, but 32 participants would only pay €3.75 per person. This is realistic, but somehow artificial. What can we do about it? For starters, we replace the assumed sports club by the class itself, who may be going on a planned trip (field trip or excursion). Next, we consider for what reason additional people might join us. One option would be that additional students (from another class of the same grade, or due to special circumstances two brothers or sisters of a student) join us on our trip. Or maybe additional supervisors are needed. After taking our proposal into account, the problem becomes:

> Our trip to the National Park is scheduled for next Wednesday. The bus will cost €120 to hire. If the bus fare is to be shared equally then how much (in euros) will each person have to pay? Maria's parents have decided, at the last minute, to volunteer as extra supervisors. How does this impact everyone's fare?

The changes to the text are rather minor, but they turn a hypothetical situation into a matter that is of immediate concern to the class. Possibly, such a trip takes place without us having dealt with quadratic equations in our mathematics class recently. If the class has already studied them, this would be a great opportunity for review.

Quadratic equations are something one often encounters and this type of activity reinforces long-term learning, ensuring they are not forgotten.

Now let us look at a problem taken from geometry:

> The shade of a 4.5 metre high tree measures 6 metres. At which angle is the sun in the sky or in other words at which angle, α, do the rays of the sun hit the ground?
>
> (Dorner, Grade 5, Problem 11c, p. 139)

We would like to turn this problem into a more realistic one and ask ourselves: Who might be interested in the outcome of this problem, and for what reason? This prepares the ground for many other versions:

> **Version 1**: The Robinson family lives at the edge of a forest. The trees south of their garden are up to 10 metres high and cast a shadow on their garden. How far away from the trees should the Robinson family plant their tomatoes, so that they can enjoy full sunlight all of the time?

We omitted some information on purpose here such as at what angle do the trees cast their shadow? There was a deeper purpose behind this omission. Students are supposed to realise that data is not always presented in a way that allows for immediate calculation of a solution. In this case they can make an estimate or do even better by learning a little about geography. The geography of the situation indicates that the angle depends on the position of the sun and hence the geographic latitude. If your students made it that far, they might also notice that the requirement of 'full sun' for the entire day does not only concern the uppermost position of the sun during summer solstice, but also for the entire growth period of the tomatoes. This requires additional information and estimates. A simple problem that requires us to calculate an angle suddenly turns into a little project on gardening and geography.

At this point the problem has become rather extensive and we considered omitting it. Instead, we reconsidered as we wanted to illustrate how slight changes to the wording, or small approximations to reality, can prompt interesting and unintended questions that are not so easy to answer. Even if you feel that you need mathematical algorithms or specific data in order to answer these questions, which your students cannot access yet without great effort, you can still open up a fruitful discussion in class by asking questions along the following lines: What did we want to know? What have we achieved? What information is missing? What would we have to do or know (as far as we can currently tell) in order to obtain a better or complete result? In some cases, what has been achieved so far is enough for putting something into practice, even if it is incomplete from a mathematical point of view. For example, if the shape of the Robinson family garden is such that only 10 m separate the fence from the house, the advice concerning the tomatoes is simple: Plant them close to the house—even if that means that they get some shade from the trees in the evening or in late summer.

Here is another observation. In most reality-based problems the question of accuracy arises. The determining factor here is the intended purpose of the solution related to the situation at hand. When something is to be planted in the garden, it is not the technical capability of our calculator that determines how many digits after the decimal point we use, but rather a thoughtful look at the garden. Is it even possible to plant tomatoes, for example, with more precision than 1 cm? Is a level of accuracy of 10 cm sufficient?

At this stage we would like to point out that sometimes our aim for a realistic solution also goes hand in hand with a more relaxed approach to mathematical completeness. Some conventional calculations of extrema for grades 9–12 (14–17 year olds) can already be solved in grades 5–8 (10–13 year olds) without knowledge of calculus, provided the aim is only to find a feasible solution.

We will rephrase the problem once more resulting in a simpler version:

Version 2: The Robinson family lives at the edge of a forest. The trees south of their garden are up to 10 metres high and cast a shadow on their garden. How far away from the trees should the Robinson family plan their outdoor seating area if they want to sit in the shade at noon even in midsummer, without having to buy a parasol?

By turning the question around, the problem becomes much easier to solve: how far does the shade reach when the sun is at its uppermost position during summer solstice? Concerning accuracy, even an estimate of about 50° for the angle the sun makes with the ground would suffice. If one person wants to sit in the shade, it would maybe make our considerations a little more complicated. In this instance we need to consider questions such as: What is the height (from the top of his/her head) of a seated individual who is sitting 5 m away from the fence? The financial incentive (i.e. not having to buy a parasol) might add a touch of reality and motivation to the problem.

Another version of the problem might be as follows:

Version 3: Mrs. Robinson loves lilacs. Therefore, she would love to plant a lilac bush in her garden, which can grow up to 4 metres tall. How far away from their property line does she have to plant the bush, so that their neighbour to the east does not complain about the shade of the lilac coming across their 1.5 metre boundary fence during the summer?

To be able to solve this problem, a few assumptions or considerations which we think your students are capable of making may prove useful. The lilac bush will have grown to its highest point by the final day of summer. This is also when the sun is at its lowest in the west. Unless the sun is covered by the house of the neighbour to the west, the shade of the lilac bush is certain to reach across the fence to the neighbour's property in the east, provided it is nearly 4 m high. How can we proceed knowing this additional information? We suggest inviting your students to find a feasible solution. Maybe the neighbour likes lilacs and he is pleased if the bush is planted close to the boundary. Maybe he just does not want any additional shade at the height of summer, as he has a vegetable patch close to the boundary. In that case, perhaps the height of the bush is up for negotiation. Is this still mathematics education? Yes, it certainly is! After all, a realistic goal is to solve real-life problems with the help of mathematics. If the result was that we could not find a solution to meet a requirement under certain conditions, the calculation indicates how essential communication is, and how important negotiations are to find a setting that makes compromise possible. What is important in real life is not doing as many calculations as possible to practice a certain algorithm but instead finding an acceptable solution.

We would like to close this chapter with some more suggested readings. For English language readers we suggest visiting:

- http://www.bowlandmaths.org.uk/
- https://www.ncetm.org.uk/resources/teaching-resources
- https://nrich.maths.org/teachers

Additionally, if you would like to learn more about converting standard textbook problems into authentic mathematics problems, you can find relevant examples on YouTube, such as the following by Dan Meyer:

- https://www.youtube.com/watch?v=NWUFjb8w9Ps

While German language readers will find additional material at:

- http://www.mister-mueller.de/mathe/unterrichtsmaterialien/

This chapter, along with the resources listed above, has highlighted several approaches to adapting textbook problems that will serve as an accessible entry point to reality-based mathematics education for teachers. In the next two chapters the authors invite teachers to engage with several more selected examples of richer modelling tasks, which lead teachers from simpler to more challenging classroom activities.

References

Baruk, S. (1998). *L'âge du capitaine (Points sciences)*. Paris: Éditions du Seuil.

De Corte, E., Verschaffel, L., & Greer, B. (2000). Connecting mathematics problem solving to the real world. In *Proceedings of the International Conference on Mathematics Education into the 21st Century: Mathematics for Living*, Jordan, November 2000 (pp. 66–73).

Dorner, E. (2006). *Elemente der Mathematik*. Wien: AHS Schulbuch.

Pehkonen, E., Naveri, L., & Laine, A. (2013). On teaching problem solving in schools. *CEPS Journal, 3*(4), 9–23.

Chapter 4
Tasks Derived from Everyday Occurrences

The aim of this chapter is to provide teachers with ideas for creating short classroom units using everyday occurrences. However, there is one characteristic of everyday life that is very challenging for mathematics teachers. Issues that arise in everyday life do not present themselves in neat bundles ready for treatment by techniques from a selected mathematical domain, but instead demands crossover between mathematical and extra-mathematical domains. We do this by creating our own mathematical models and sometimes working with other peoples' models. We have encouraged you to work with us in this way in previous chapters. Very often the real occurrences are already described in print and/or diagrams as is the case in newspapers, drawings, plans and graphs of various kinds. This means we have to work with other peoples' representations or models and knowing how to do this greatly improves our mathematical competence. It is important that teachers are able to engage in this change of perspective and carry it through into their classrooms.

In the last chapter we considered textbook problems, and tried to see how realistic they were and how we could make them more up-to-date, relevant and interesting. We gave you plenty of suggestions on how this can be achieved. In this chapter, we will start from a different perspective, by creating small classroom units using everyday occurrences. Frequently, these occurrences are newspaper or magazine reports, situations taken from different professions or everyday life, or events from nature or technology that we, or other people, notice. In some cases, we also use projects from our own students as a starting point (whilst acknowledging these students for their contribution).

This different approach has one important consequence for the proposed classroom materials. In our previous chapter we used problems that could clearly be linked to a particular mathematical topic. In this chapter we select topics from newspapers or our professional and everyday experiences, inspired by our curiosity and a desire to better understand the world. As a result of this, our problems cannot always be linked to one specific area of the syllabus or a particular educational level. From experience, we know that in many cases various mathematical methods have to be employed to solve a problem. When we take a closer look at a chart in a magazine, for example, we need some knowledge of descriptive statistics. In the course of this analysis we might also encounter a problem that can be solved using linear

J. Maaß et al., *Mathematical Modelling for Teachers*, Springer Texts in Education, https://doi.org/10.1007/978-3-030-00431-6_4

equations or geometric analysis. This approach is atypical of classroom instruction. Thinking and practicing within the boundaries of labelled mathematical domains is much easier because we know what to expect.

From the perspective of didactics of mathematics, this comprehensive approach to making connections between different mathematical domains is what makes long-term mastery of mathematics possible. The Society of Didactics of Mathematics has created a study group for this purpose. In the preface to the first three volumes of the series 'Mathematical Networks', containing classroom materials with a variety of mathematical problems, the authors write (and translated here):

> In this project our main aim is to provide students with a better understanding of mathematics and with more motivation to study mathematics. We would also like them to better understand how the various domains of mathematics relate to each other and our everyday world. Currently mathematics education often means neatly dividing mathematics into its domains, which are to be studied independently until the next test and can easily be forgotten again afterwards. As opposed to this, we perceive studying and teaching mathematics as a network of interrelations, a cross-linked body of knowledge leading to meaningful long-term learning.
>
> (Brandl et al. 2013, p. 1).

Is there an additional requirement for the effective teaching of mathematics? What about exploring the links between mathematics and the real world to make mathematics more useful, interesting and accessible? For further details on this, please read the arguments of Maaß and Wildt (2012).

Before presenting you with examples, we would like to stress that it takes effort and a thorough understanding of mathematics to identify topics in everyday life, such as media reports, which can be developed into applicable mathematics lessons. We call it a 'mathematician's eye' (see the work of Terry Maguire at http://www.haveyougotmathseyes.com/ or Androsch et al. 2015). The ability to view the real world using a mathematician's eye is something that can easily be developed. After developing this ability, plenty of potential learning opportunities present themselves.

The aim of this chapter is to provide you with (plenty of) examples of everyday occurrences that have been adapted for use in the classroom, and also to assist you in developing your own problems. Therefore, we will maintain our interactive style and frequently ask questions, invite your active participation, and request you to work independently. Here we also stress the need for teachers to encourage their students to formulate their own problems and collect their own data.

4.1 Ramp It Up

This idea is credited to Mrs. L. Huber who prepared a mathematics course unit for a workshop on teaching mathematics in Linz, Austria in the summer of 2014. We acknowledge her contribution with thanks.

Overview: This problem focuses on the living arrangements of a person in a wheelchair who lives in a block of flats. After entering the front door of the building, two steps have to be negotiated before the lift can be accessed. A new wheelchair ramp is required to provide easier access. By using our knowledge of trigonometry and gathering information on wheelchair ramps, we should be in a better position to decide which wheelchair ramp is most appropriate.

How would you present this problem to your students?
Please jot down some ideas so that you can compare your suggestions to ours.

Implementation: We suggest inventing a story around the problem. For example:

Our school has made contact with a school in Malmö, Sweden to organise an international student exchange program. The Swedish students are scheduled to stay with us for four weeks, and vice versa, we get to stay with the Swedish students. Class 1B has been chosen by the school principal for the student exchange. One Swedish student, Lars, is in a wheelchair. He is to stay with the Ryan family. The problem is that the Ryan family lives on the second floor of an apartment block. The lift in this building only goes as far as the concourse/reception and after this there is a staircase, consisting of two steps, down to the front entrance. It is simply impossible to mount the two steps in a wheelchair without assistance. Therefore, Class 1B has suggested the idea of installing a wheelchair access ramp to overcome this problem. However, they are not able to calculate the required length and angle of the ramp. Can you help Class 1B resolve this problem?

Instructions for the class:

- Form groups of three or four students.
- Consider what information you need to help Class 1B determine the required length and angle of the ramp.
- The results will be gathered on the board after 10 min.

Here are some questions that you could pose to assist your students if they are struggling to generate ideas:

1. What does the staircase look like?
2. Is there enough space for a ramp?
3. Which types of ramps are available?
4. Where could we purchase these from?
5. What is the maximum possible angle for the ramp so that it is suitable for a wheelchair?

Some suggested responses to these questions are provided below, although in an ideal scenario students should collect their own data to increase the authenticity of the problem.

Q1: On entering the building you will find the letterboxes on the right-hand wall, and immediately behind them is the staircase. This layout means there is enough space for a ramp longer than the staircase itself. The steps are 18.5 cm high, 120 cm wide and 30 cm deep.

Q2: We have come across two suppliers on the web:

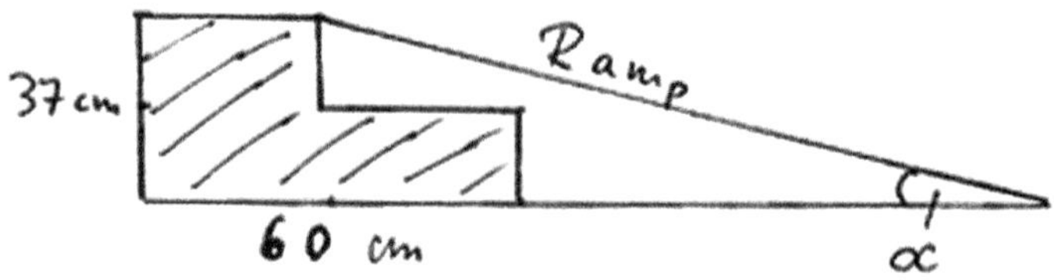

Fig. 4.1 Ramp dimensions

1. Pro Mobility offer a telescopic ramp

   https://www.promobility.ie/access-ramps-2/

2. Roll-A-Ramp offers a rollable ramp

   http://www.rollaramp.com/

Each ramp makes use of different technology, therefore each has its pros and cons.

Q3: Wheelchair access guidelines generally recommend a maximum gradient of 8.33% for wheelchair ramps. This is equivalent to a 1 unit vertical increase for every 12 unit horizontal increase (1:12).

For more information on wheelchair ramp guidelines visit the following websites:

http://www.wheelchair-ramps.co.uk/information/
http://www.adawheelchairramps.com/wheelchair-ramps/ada-guidelines.aspx

Hence, to overcome two steps which are 37 cm high in total would require a ramp with a horizontal distance of 4.44 m.

Having collected this information, we can now engage in further mathematical tasks. With the general educational goal of individual responsibility in mind, we suggest having your students plan the next steps themselves. To help your students, you could pose the following questions:

- How should we proceed from here?
- What would you do next?

We suggest drawing a diagram to represent the information available. Please note there are better and more precise ways of sketching the problem than shown in Fig. 4.1, especially if we use scale paper or technology.

We will now consider this problem in terms of cost and practicality. But first we have some questions for you, the teacher, to consider.

> Will a diagram like that shown in Fig. 4.1 suffice for this purpose?
> Does it contain all the essential details?
> Are we now in a position to start our calculations?

Unlike textbook problems, the steps required to determine the cost in this case are not prescribed and we suggest encouraging your students to find out for themselves what has to be done next. One option could involve us using the maximum gradient of the ramp to calculate the maximum angle, α, in degrees.

Using the tangent ratio, the horizontal distance from the bottom step to the end of the ramp can be calculated. Using this information, and referring back to the two different suppliers' websites, the feasibility and cost of the ramp can be determined.

Note

To convert between % gradient and degrees, you can use the following formulae:

$$\% \text{ gradient} = 100 * \left(\frac{Y}{X}\right) \text{ (vertical rise / horizontal run)}$$

$$\text{degrees} = \tan^{-1}\left(\frac{Y}{X}\right)$$

There is, however, a more direct approach that we could use to achieve our goal. If two steps, measuring 37 cm in height, are to be scaled with a maximum gradient of 8.33%, this means an increase in height of 8.33 cm for every horizontal metre. Therefore, the ramp has to be longer than 4.44 m so that Lars can ascend it without help. The ramp measuring a little over 2.4 m offered by Pro-Mobility will not suffice in this case. The 'roll-a-ramp' portable ramp costs 105 dollars per foot which would mean that a ramp measuring just over 4.44 m would cost approximately €1,440 plus VAT and shipping (1 Euro = 1.062 USD)—a lot of money for a four week student exchange project!

We therefore suggest that someone accompanies Lars to provide him with assistance when using the ramp. To determine the feasibility of this, we suggest having your students carry out another calculation first. For example, what would the gradient be if we assumed that the school could afford a ramp that is 2 m long? If a height of 37 cm has to be negotiated over a horizontal distance of 2 m, this means a gradient of 18.5%. In this case someone in a wheelchair would definitely appreciate a helping hand!

Comment: The main challenge in this example is to identify the underlying problem. Having done so, the calculations are relatively straightforward. A quick calculation showed that a ramp with a gradient of 8.33% was much too expensive. In general, if a question on whether or not something makes sense arises in real life and we can answer it by employing mathematics, as was the case here, the merit of applicable mathematics education immediately becomes apparent.

4.2 Which Road Should We Take?

This idea is credited to Mrs. M. Spiegl and Mr. T. Lehner who prepared a mathematics course unit for a workshop on teaching mathematics in Linz, Austria in the summer term of 2014. We acknowledge their contribution with thanks.

Overview: When using a satellite navigation system you will often be provided with more than one possible route. Mathematics can help us determine the most suitable route. If we assume that we are not just looking for the shortest driving time, then other factors will need to be taken into consideration such as shopping on the way, paying Aunt Anna a visit, enjoying the scenic view, or avoiding tolls. When such considerations are taken in account our deliberations might become more complex.

> How would you present this problem to your students?
> Please jot down some ideas so that you can compare your suggestions to ours.

Implementation: We suggest inventing a story around the problem. For example:

Your geography teacher is organising a field trip from your school (Europagymnasium Auhof, Linz) via *bus to the abandoned opencast pit in Eisenerz, once the principal centre of Austrian Iron mining. The bus operator enquired which route we would like to take. Route planners (such as Google Maps or Open Street Map) provide us with three suggestions:*

1. *The first route takes us south* via *the A1 motorway passing the city of Steyr.*
2. *The second route is* via *the A9 motorway and then takes us east.*
3. *The third route takes us southeast passing the city of St Valentin.*

A rough sketch of the three suggested routes are shown in Fig. 4.2. Using Google Maps, or another web-based mapping service, construct a detailed description of the route, outlining travel time, distance travelled and roads traversed. What considerations must be taken into account when you compare these three routes?

We propose a brainstorming session with the entire class. After studying the map, what suggestions have the class come up with concerning the three routes? Here are our remarks:

- Route 1 (via A7/A1): According to Google Maps the route is 141 km long. It takes a car 2 h and 8 min to travel this route and you have to pay tolls on the highways. We have to ask the bus driver if he has an electronic pass for the tolls. According to traffic reports there is a construction site on the national road 115, north of Landl, possibly resulting in delays. If this is the route to be taken, we have to check the status of the construction site once again before departure. Approximately 64% of the route is on mountain roads with only 18% on highways.
- Route 2 (via A9): According to Google Maps the route is 163 km long. It takes a car 2 h and 8 min to travel this route. The largest proportion of this route is via motorway. In the Ennstal area we would go down national road 146, and towards

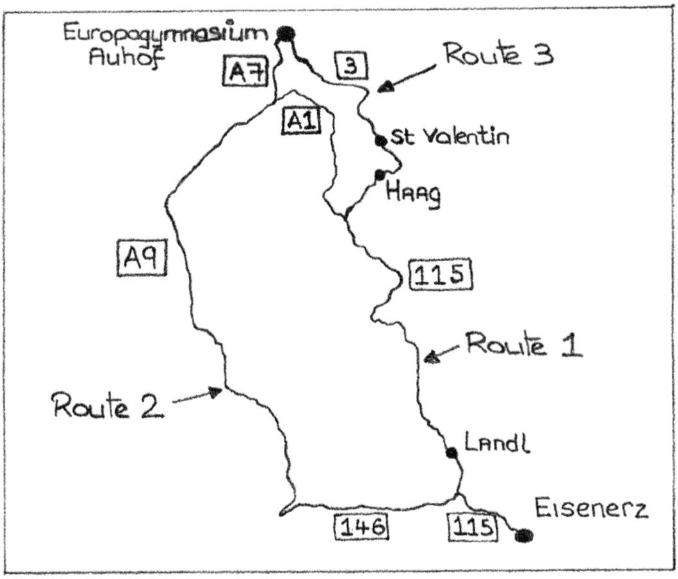

Fig. 4.2 Map of three possible routes

the end travel for a few kilometres on national road 115. On this route we have to pay a toll for using the Bosruck tunnel. We have to ask the bus operator if he has an annual pass for the tunnel or if we have to pay for using the tunnel.

- Route 3 (via Haag): According to Google Maps the route is 147 km long. It takes a car 2 h and 15 min to travel this route. Initially we proceed eastwards along the national road 3 before turning south to join the A1, heading southeast for a few kilometres. We then merge onto the national road 42 and head south again to join up with the national road 115.

Do we have enough information now to make an informed decision? We would like to point out that everyday decisions regarding what route to take are generally made without much deliberation. The level of detail outlined above is generally sufficient as the difference in time and cost between routes is minimal once no additional reasons for changing our route arise. However, if one does not simply take the information on the internet for granted you might notice that the suggested driving time is for cars. Buses on the other hand might take longer on these roads. In this case we need to put more thought into our decision.

We would like to point out that, as in most modelling scenarios, we need to decide if we believe the information provided by sources such as the internet, advertisements, companies or public authorities. Mathematics can be a means of verifying such information and checking if it makes sense. However, this requires some time and effort on our part. We consider it very important for students to realise in school that situations requiring such decisions do frequently arise in real life. In this book we advocate the teaching of applicable mathematics in a way that students can use it to verify facts they are not entirely convinced of.

How can we verify the information provided by Google Maps? Below we have listed possible options that you could consider:

1. Use a different satellite navigation system.
2. Use a map and calculate the distance the old-fashioned way (using pins and a string; this is a good way of practicing calculations with map scales).
3. Accept the proposed distance but calculate the driving time yourself.

Here we shall only focus on option 3. Students should realise themselves that they need information on the average speed a bus travels down a motorway, or on national roads in the plains or in the mountains. To assist them in this endeavour they could call the bus operator or make their own estimates.

We will now carry out a short modelling task using estimated values (or values deemed realistic for buses by automobile associations in Austria):

- average speed on the motorway: 80 km/h
- average speed on national roads in the plains: 60 km/h
- average speed on national roads in the mountains: 40 km/h

Bear in mind, as with all modelling problems we performed, we can revisit these assumptions later on or carry out further research on them and recalculate how these changes influence our results (i.e. driving time). Depending on the model, it might be advantageous to see what effect changes made to the initial values such as average speeds have on the desired value to be calculated such as total travel time, and if this can be expressed as a function.

Using our assumed values, route 2 consists of 109 km of motorway, which takes 1 h and 22 min to travel (provided there is no traffic jam or no queue for the toll, as the bus might be able to use an electronic toll system). We will then travel down the Ennstal national road for approximately 38 km (38 min) and then travel another 16 km on national road 115 through the mountains (24 min). According to our calculations it will take a bus approximately 2 h 24 min; 16 min slower than the time it would take a car, according to Google Maps.

How accurate was your calculation of the driving time? What degree of accuracy makes sense in this case? Our calculation was kept rather inaccurate on purpose, as we only wanted to re-confirm our assumption that a bus would take longer, on average, than the times Google Maps suggested for a car.

We calculated the driving time for the other routes in a similar fashion. For route 1 the result was approximately 3 h, and for route 3 approximately 3 h 10 min. Therefore, if driving time is the decisive factor, route 2 is the best option.

As we have just seen, driving time is influenced by the approximations of the average speed that the bus can travel. If your class requires more information on this, we suggest analysing the various driving times using a spreadsheet. We have chosen to use a spreadsheet as it will allow us to perform accurate calculations whilst also allowing us to make quick changes and immediately see the revised results. Sample spreadsheet outputs are given in Table 4.1.

Table 4.1 Spreadsheet data for calculating total driving time

Driving time: Linz - Erzberg		
Assumed average speed of a bus		
Motorway	80 km/h	
National road in the plains	60 km/h	
National road in the mountains	40 km/h	
Route 1 (via A7/A1):	**Distance**	**Driving Time**
Motorway	26 km	19.5 min
National road in the plains	24 km	24 min
National road in the mountains	91 km	136.5 min
Total Distance	141 km	180 min
Route 2 (via A9):	**Distance**	**Driving Time**
Motorway	109 km	81.75 min
National road in the plains	38 km	38 min
National road in the mountains	16 km	24 min
Total Distance	163 km	143.75 min
Route 3 (via Haag):	**Distance**	**Driving Time**
Motorway	9 km	6.75 min
National road in the plains	47 km	47 min
National road in the mountains	91 km	136.5 min
Total Distance	147 km	183.50 min

Using the spreadsheet it would be relatively straightforward to calculate how long the journey would take on an express bus that can travel at increased speeds. By simply altering the average speed at which the bus can now travel our total travel time will be automatically updated. Given that the average speed of the express bus is 100 km/h on motorway, 80 km/h on national roads in the plains, and 60 km/h in the mountains, the total travel time for each route is recomputed in Table 4.2.

As you can see, these results are even faster than those indicated by Google Maps. We encourage you to discuss this point with your students. Why do they think this is the case? What factors could be affecting the travel times suggested by Google Maps? Will the travel times suggested by Google Maps always remain the same?

Finally our class will have to make a decision on which route to take. Travel time and cost will probably be the two key factors that our class will need to consider before reaching a decision. Due to relatively small differences in travel time, the decisive factor will probably be whether or not the tolls on the A1, A7 and A9 motorway will incur extra cost, and how much that cost will be.

Comment: We have located this example in Upper Austria, as these three routes are relatively close in both distance and travel time, with each having its pros and cons. It could be an interesting task for your students to find a similar example closer to home with routes that are comparable as we have done in our example. Without

Table 4.2 Data for calculating total driving time for express bus

Driving time: Linz - Erzberg		
Assumed average speed of a bus		
Motorway	100 km/h	
National road in the plains	80 km/h	
National road in the mountains	60 km/h	
Route 1 (via A7/A1):	**Distance**	**Driving Time**
Motorway	26 km	15.6 min
National road in the plains	24 km	18 min
National road in the mountains	91 km	91 min
Total Distance	141 km	124.6 min
Route 2 (via A9):	**Distance**	**Driving Time**
Motorway	109 km	65.4 min
National road in the plains	38 km	28.5 min
National road in the mountains	16 km	16 min
Total Distance	163 km	109.9 min
Route 3 (via Haag):	**Distance**	**Driving Time**
Motorway	9 km	5.4 min
National road in the plains	47 km	35.25 min
National road in the mountains	91 km	91 min
Total distance	147 km	131.65 min

localisation, the problem will lose its appeal. This insight holds true of all problems incorporating reality. If the reality to be modelled is too far removed, it is not as motivating as local problems. Therefore, it is often up to the teachers to adapt a proposed lesson based on reality to the living environment of their students and let them find up-to-date data that is meaningful to them.

4.3 Beverages for the Parent-Teacher Evening/School Party

Overview: A class is responsible for the sale of beverages at the upcoming school disco. They wish to make a profit from this venture but to do so they must first carry out some basic calculations.

> How would you present this problem to your students?
> Please jot down some ideas so that you can compare your suggestions to ours.

Implementation: We suggest inventing a story around the problem. For example:

The principal of your school announces that your class will be responsible for the organisation and running of the upcoming school disco, scheduled to take place next month. She suggests that this may be an opportunity for the class to make a small profit by selling a variety of beverages. The principal asks for volunteers, some to help with the running of the disco on the night and others to help with organising the drinks stall. The latter group will be responsible for deciding which drinks will be sold and for how much?

After a class brainstorming session, it may be suggested that the beverages be grouped into hot and cold beverages: tea and coffee versus mineral water, fizzy drinks and fruit juice. Next, teams responsible for hot drinks and cold drinks will be formed and they will consider the following questions before proposing a price structure to the entire group:

- How many drinks will we need in total?
- Who will do the shopping?
- How many disposable cups will we need?
- How much do they cost?
- Will the school charge us for boiling water? Electricity: flat rate or per kilowatt-hour (kW h)?
- Once all costs have been accounted for—how much should we charge for the beverages?
- What should we do with the leftover beverages?
- Should parents/teachers who supervise on the night also be charged for drinks?
- Will we do the cleaning up and waste disposal ourselves?

For teachers experienced in organising events such as school discos, answering your students' questions should be easy. The tasks of planning, shopping, and pricing should be relatively straightforward for someone with experience of running an event like this. Although it is much more efficient for you to share your experiences with your students so that they can arrive at a solution more promptly, letting them make their own assumptions, conduct their own research and allowing them to plan everything, would be more educationally valuable. Decisions regarding how much we should scaffold our students are a fundamental challenge for us as teachers, but also present us with great opportunities.

In conventional mathematics, results calculated and problems solved mostly serve the purpose of answering the question: is the result correct? It is not that important to students whether the virtual price of a cup of mineral water is €1 or €10. What matters more to them is to calculate correctly how much profit would be generated if they sold a certain number of cups, assuming the cost price is 50% of the sales price. On the other hand, our situation involves dealing with actual money in a real life scenario which would better serve to engage students. When the problem is presented like this, students are not likely to suggest €10 per cup of mineral water, nor will they assume that the cost price is 50% of the selling price or suggest buying 100 L of mineral water. You as a teacher might be surprised by how rational and focused on reality your students can be when it comes to planning and making decisions that affect them. All it takes is for them to realise that their involvement is essential, that they will be held responsible for their decisions, and that they will benefit from their calculations.

Comment: We do not intend to try your patience here by calculating how much the class will earn per cup if they purchase one litre of mineral water for 30 cents and then sell it for €1. In the case of coffee, the calculations become more interesting if you compare a cup brewed in an old-fashioned coffeemaker using a conventional filter, to a cup pressed from a capsule.

At this point we would like to stress that similar modelling problems can be derived for other types of food and drink, such as home-made cakes. Provided facilities like a school kitchen are available, and your cookbook contains a recipe for a particular cake, you can also calculate ratios to find out how much flour or how many eggs are required for 5 cakes?

This example also affords us the opportunity to answer an important question; what happens if students make the wrong decision(s)? In this case, what if too much or too little coffee or mineral water is purchased? If such a mistake was made while solving a textbook problem, students might get corrected, scolded, or receive a bad grade. In our scenario some students at the disco might get irritated by the lack of drinks available and/or less profit might be generated. This is why we think students, in this case, are more likely to learn from their mistakes. Moreover, it is interesting to find out, together with the class, who is responsible for the mistake or miscalculation, and how it occurred.

4.4 Furnishing a Room

Overview: After moving house Kate's parents have decided to allow her to furnish her new room with her existing furniture and/or a small selection of new furniture. The layout of the room and the furniture available (or the available budget) is already specified. Our goal is to help Kate furnish the room in a pleasant and efficient manner.

> How would you present this problem to your students?
> Please jot down some ideas so that you can compare your suggestions to ours.

Implementation: Let us start with a simple version, whereby the available furniture is known, and the layout is defined, as shown in Fig. 4.3.

Table 4.3 provides a list of available furniture. It is not necessary to use all items.

The following are some possible instructions you could give to your students to get them started:

Kate gets to decide how she will furnish her room after the move. She has a selection of furniture at her disposal as outlined in Table 4.3. A bed, a wardrobe and a desk will be essential pieces of furniture. Form small groups and come up with a proposed layout for the furniture in her new room. During the next class period all

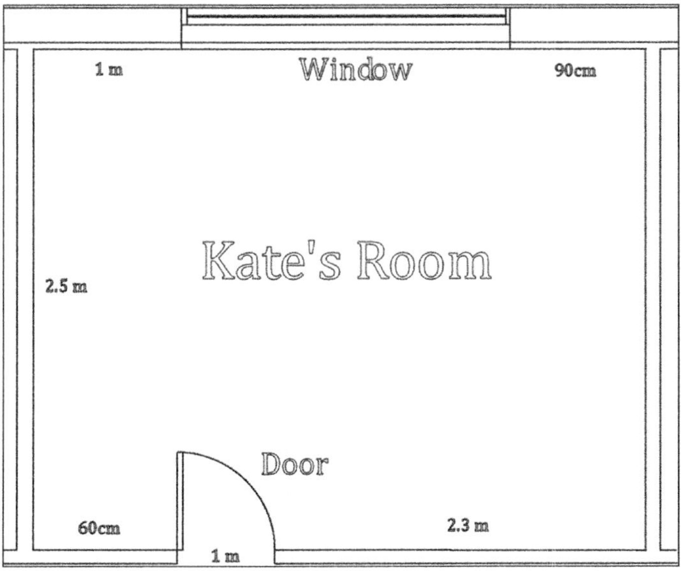

Fig. 4.3 Room plan for Kate's room

Table 4.3 Available furniture for Kate's room

Bed (1 m wide, 2 m long)	Bedside table (60 cm long, 60 cm deep)	Wardrobe (1.2 m long, 50 cm deep)
Chest of drawers (1.2 m long, 40 cm deep)	Desk (1.6 m long, 70 cm deep)	Table (80 cm long, 80 cm deep)
3 chairs (60 cm × 60 cm each)	Shelf (1 m long, 30 cm deep)	1 armchair (70 cm × 70 cm)

proposals will be presented and the class will choose what they believe is the best layout for the furniture.

Another possible suggestion is:

For each item of furniture draw the layout on a separate sheet of paper. Consider where to put each item of furniture so that the room looks nice and is fit for purpose.

Comment: On the one hand more items of furniture are available than can be fitted into the room and on the other hand more than one solution is feasible. Therefore, the selection of the best layout should be a two-step process. To begin, we should check for functionality (can the doors of the wardrobe, the room door, and the window be fully opened without obstruction? or Will Kate stumble over the chair on her way to the bed?) If all groups answer "Yes" to these questions then the decision regarding the best layout becomes more difficult as the remaining proposals will be compared solely on their aesthetic appeal. When functionality is an issue, the teacher may offer some suggestions, but decisions relating to the aesthetic appeal should be left to the students.

Obviously this task can become more complex, for example if the size of the room is increased, or if there are more furniture specifications provided, or if there are more furniture options available. Extensions to this problem could include decorating an L-shaped room, a gazebo, or even an entire apartment.

4.5 Mice in the Granary

This problem arose following discussions between one of the authors (JM) and Dr. Gunther Ossimitz. For more on strategies for teaching system dynamics see the following link: https://www.systemdynamics.org/bibliography.

Overview: Mice in a granary are a good example of a dynamical system suitable for modelling. This type of problem also enables students to gain an insight into dynamical ecological systems. At first, the mouse population increases exponentially. Later on the population increase does not follow a logistical growth pattern as you might expect, but instead catastrophe strikes. From the perspective of the mice, the excess population in only one room (the granary) is so threatening that they stop procreating and instead kill each other. Only after their numbers have decreased significantly, will they revert to normal behaviour and start procreating once again.

How would you create a problem using this information?
Please jot down some ideas so that you can compare your suggestions to ours.

Implementation: We convert this biological information regarding the behaviour of mice into a story, resulting in the following problem statement:

Farmer Jack Connors complains about losing a significant portion of his grain due to mice. Occasionally there appears to be no mice at all in his granary. At other times, he notices that there are a very large number of mice present, and from time to time he is astonished by the large numbers of dead mice that he finds. Eventually, to address the problem of the mice, he buys some poison. However, the poison does not kill all the mice. It only kills approximately 50 mice. Eventually he wonders: Is there an optimum moment for using the poison? Farmer Connor's sister is a biology teacher. She vaguely remembers that the mice population will increase by 10% per month, provided there is enough feed and space. Furthermore, she knows the reason why there are so many dead mice from time to time. From the perspective of the mice, the excess population in the granary is so threatening that they stop procreating and kill each other instead. Only after their numbers have decreased to approximately 5% of the population before the crisis, will they revert to normal behaviour and start procreating once again.

Having presented this story to your students their task would then be to create a mathematical model that describes the mice population in the granary at any given time. We suggest starting with 100 mice for their first model, and assume the catastrophe strikes when the population reaches 10 times the initial value. Is it possible to tell from the model when would the best time be for the farmer to spread the poison?

When faced with this task, does a system of differential equations come to mind? You might consider this to be far too complex! In that case we have a pleasant surprise for you: this problem has already been solved by students as young as 10 years old using spreadsheet analyses. To do this we first need to gather the available information and construct a model using a spreadsheet program.

Available Information:

- Initial population: 100 mice.
- Normal growth rate: 10% per month.
- Maximum population: 10 times the initial population value.
- Population after systemic crisis: 5% of the population before the crisis.
- Using poison kills off at most 50 mice.

We created the following spreadsheet to represent the mice population in any given month.

	A	B
1	**Month**	**Number**
2	0	100.00
3	1	110.00
4	2	121.00
5	3	133.10
6	4	146.41

Eventually, after 25 months, the population surpasses 1000 mice. At this point we would ask students to calculate 5% of this population value which is the new population of mice for the subsequent month.

26	24	984.97
27	25	1083.47
28	26	54.17
29	27	59.59
30	28	65.55

Afterwards, the growth resumes at 10% until the number of mice surpasses 1000 again:

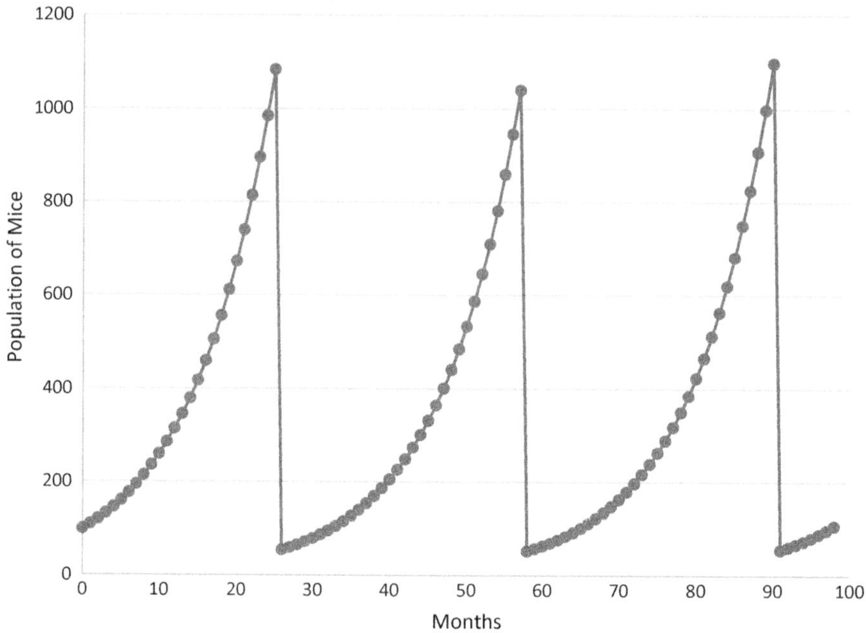

Fig. 4.4 First simple model of the population of mice

58	56	945.30
59	57	1039.83
60	58	51.99
61	59	57.19
62	60	62.91

Once this data has been generated, we can easily represent it on a graph using our spreadsheet program, as shown in Fig. 4.4.

Figure 4.4 illustrates how many mice will be in the granary at any particular time over the course of the first 100 months. It certainly makes sense to study this graph in more detail and from this, attempt to draw our own conclusions regarding the optimum time to spread the poison. What did you notice from your study of Fig. 4.4?

We would like to draw your attention to the repetitive dynamics of the population of mice. Whenever the total surpasses 1000, catastrophe strikes and 95% of the mice are killed off. Then exponential growth resumes. Figure 4.4 thus illustrates what has been described in words at the outset.

By foregoing the tools of analysis, and not conducting an in-depth analysis of the problem, we miss out on the chance of calculating the number of mice in the granary at any given time to a certain degree of accuracy. This, however, is of no great consequence to us, as we only know, to a certain degree of accuracy, how many

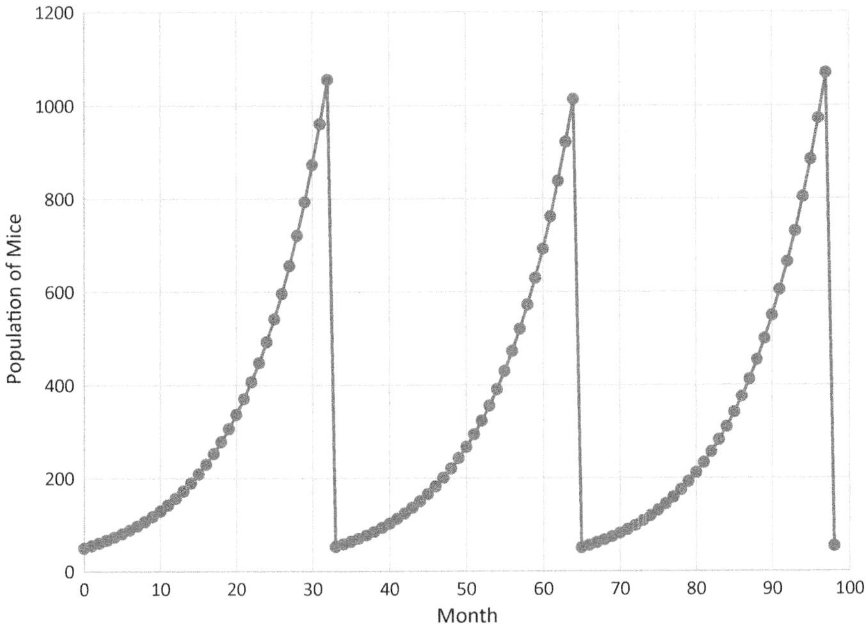

Fig. 4.5 Population of mice with poison administered on day 1

mice live in the granary at any given time anyway. If we are looking to determine the number of mice present every second, or in even smaller intervals, the result will be fractions of mice or may not change at all.

In order to find the best moment to administer the poison, a monthly interval will suffice. The question that we now must address is what month is best to administer the poison so that it has the maximum possible effect on the population of mice. We can explore our options using the spreadsheet. First subtract the number of mice killed by the poison at any given time from the overall population of mice. How does this affect the graph?

In Fig. 4.5 we administered some poison on the first day, killing off 50 of the original 100 mice. It now takes 32 months before the first catastrophe occurs, 7 months longer than the original model, where no poison was administered.

Figure 4.6 shows the case where poison was administered on the first day of month 15. In this case, it took 26 months (one additional month) for the population of mice to exceed 1,000.

In the third scenario poison was administered at the start of month 25, as shown in Fig. 4.7, with hardly any effect, as mass extinction would occur a little later anyway (the following month).

Finally, our 4th scenario, depicted in Fig. 4.8, involves the farmer administering poison immediately following a catastrophe, when the population of mice was at its

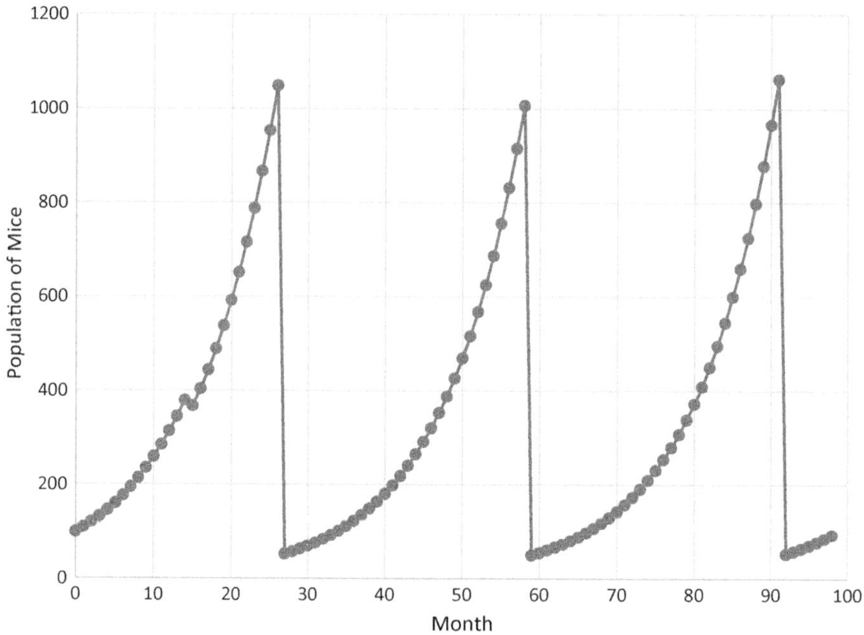

Fig. 4.6 Mice population with poison administered at the beginning of month 15

lowest. In this scenario, the next catastrophe takes place after 84 months, compared to 57 months in our original model (Fig. 4.4).

To sum up the result, we would advise the farmer to administer the poison once he notices a lot of dead mice in the granary, as this suggests a catastrophe has just occurred. However, as the poison does not kill precisely 50 mice, and some mice might migrate from somewhere else, the farmer's hopes of having a mouse free granary are probably in vain.

Comment: Clearly, there is scope for extensions to this problem. One obvious step would be to alter the initial values, for example you could assume an initial population of 60 mice, an 8% growth rate, and the value when catastrophe strikes to be 500% of the initial population of mice. In this case you will note that the overall results will be similar. Alternatively, introducing cats to the model would make it more interesting from a mathematical perspective. In this scenario we could assume that each cat eats one mouse a day, and well-fed cats will have an approximate procreation rate of 3% per month. Introducing the cats to this model has opened the door to predator–prey models. Once again our suggestions for dealing with this extension are in line with those of G. Ossimitz—it would be better to avoid the use of differential equations at this stage and instead use spreadsheet analysis with fixed time intervals.

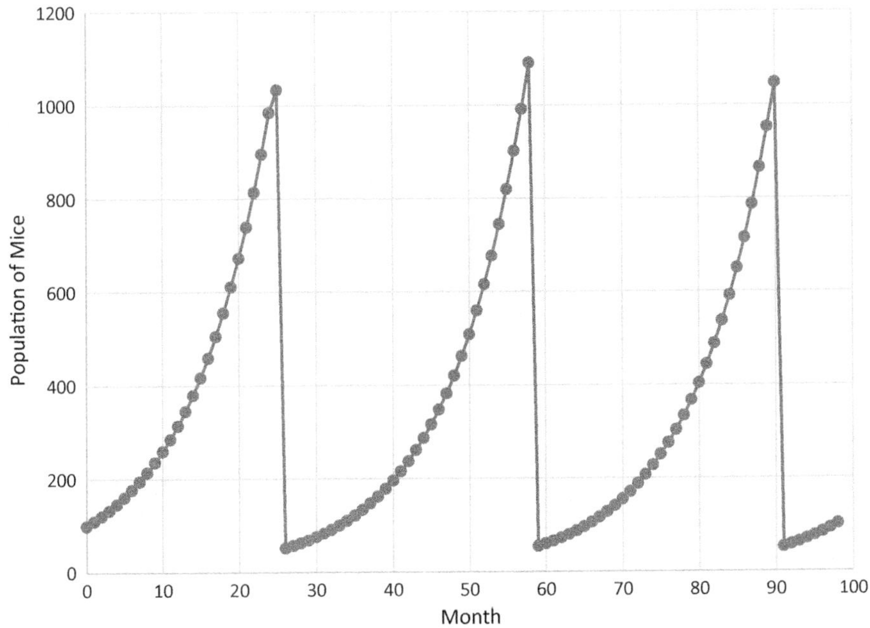

Fig. 4.7 Mice population with poison administered at the beginning of month 25

4.6 Robotic Control

The idea for the following example was developed as part of a project called 'Power Girls' (http://www.edugroup.at/praxis/portale/powergirls/projekt.html. Please note that this website can be translated into the English language).

Overview: A student who takes on the role of a robot shall be directed towards a destination within the classroom as precisely as possible by a set of instructions created in advance by other classmates. To facilitate this we need a command language (one with coordinates would be preferable). Accuracy is the main problem here: How best can we measure and standardise the step size so that the 'robot' will arrive precisely at his/her destination? Industrial robots encounter issues with accurate robotic control (such as a robotic lawnmower or a vacuum cleaning robot) and these issues need to be addressed.

How would you create a problem using this information?
Please jot down some ideas so that you can compare your suggestions to ours.

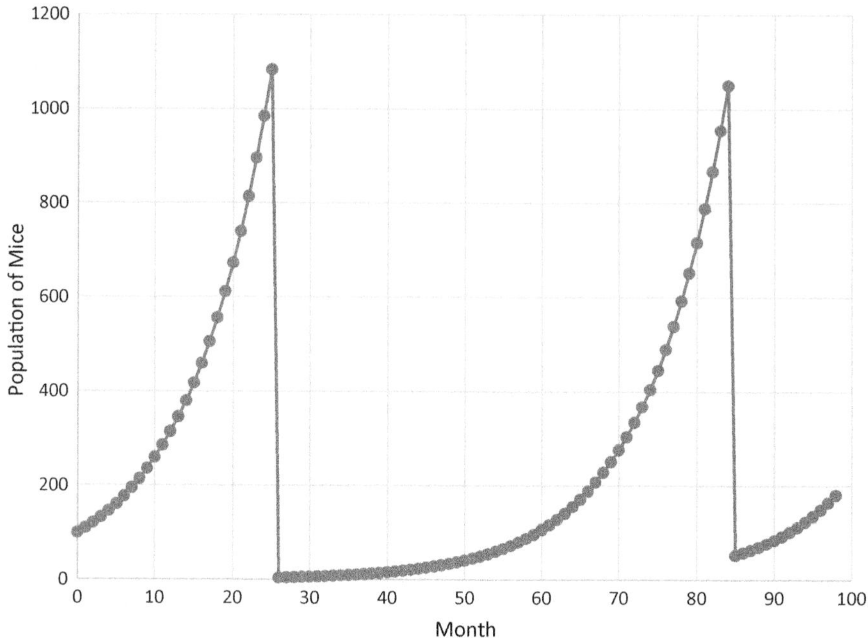

Fig. 4.8 Mice population with poison administered at the beginning of month 26

Implementation: In this section we elaborate on possible teaching methods for this problem, so that it becomes apparent how the two central educational goals of the problem (using coordinates and accuracy) can be achieved.

As an introduction, a video showing how robots can be controlled might be suitable (https://www.youtube.com/watch?v=_JRV2Z1mr7M). The question now arises: how are robots actually controlled? Somehow, the robot has to be instructed with commands, delivered in the correct sequence, if it is to get from location A (starting point) to location B (final destination). How to accurately control the movements of all robots is a problem intrinsic to study of robotics. Accurate control of movement is very important in practice. For example, in industry the weld spots have to be put in the right place so that welding can be performed. Additionally, parts that need to be assembled have to be moved to the right position so that screws can be fixed and paint applied in the appropriate places. In this classroom activity, a student gets the opportunity to become a robot in order to simulate simple robotic control.

The teacher can start by inviting students to engage with a basic robotic movement problem on the blackboard. Suppose a simple robot can only move in two directions: up and down or left to right (parallel to the edges of the blackboard, not diagonally) and can only move in one of these direction at any given time. If we want the robot to move from one point on the blackboard to another, we have to give him (the robot) precise instructions. Initially, we have to define our starting point; coordinates can be useful in this regard. We can set up our coordinate system so that our starting point is at the origin, and we assume that our goal is at the point (30, 50). We can thus describe the route (instructions) for our student (robot): Move 30 units to the right and then 50 units upwards (or vice versa).

If we want to work without coordinates, we could look for the point where the lines intersect that are parallel to the edges of our blackboard, and passing through our starting point and destination point. We could then measure the distance between our starting point and either intersecting point, and also the distance between the destination point and this intersecting point. This is the information needed to generate the commands for the robot.

Next, another preparatory step may be taken. The teacher can assign a slightly more challenging task to everybody. Suppose the robot is to move from point (0, 0) to point (4, 9). However, there is a rectangular object obstructing its path. The object's corners are located at (0, 2), (5, 2), (0, 3) and (5, 3). What instructions should you now give the robot, who can only move in unit steps, to allow him/her to reach the destination?

One possible solution might be:

1. Move to the right until you reach the point (6, 0).
2. Then move up until you reach the point (6, 9).
3. Finally, move to the left until you reach the point (4, 9).

So far, in this scenario, we have not considered the issue of accuracy. Working with the coordinate plane on graph paper or on the blackboard means accuracy is relatively easy to obtain. Students are supposed to realise the problem of accuracy themselves when controlling their robot. To achieve this aim, we engage students in a simulation.

Students should split into groups and create their own instructions for robotic control. The teacher explains the 'rules': One member of the group will act as the robot who has to follow the instructions precisely. As on paper, or on the blackboard, the robot can only move in a bi-directional way, that is up and down or left to right (parallel to the wall, assuming a rectangular classroom or otherwise draw lines or use desks to make a rectangular space).

In this simulation, one rule is particularly important: The robot cannot measure distance; it can only count steps. An additional rule is that each series of steps has to be preceded by alignment, that is, before taking a step the extended arms of the robot have to be parallel to one wall and at a right angle to the other one. Also, the robot is not allowed to turn during a single series of steps!

First, the teacher might advise students to conduct a test run: Every robot shall move from one window to the opposite wall. Students in each group must generate a set of instructions for their robot. For example move forward 15 steps or something similar. At this stage the teacher moves from group to group, monitoring cooperation within the groups and providing assistance where needed. One after the other, the robot in each group moves strictly in accordance with the instructions provided to them. Students must then mark the finishing location of their 'robot' to see how accurate each group's 'robot' was in reaching its destination. Then students should measure the distance from the finishing location of the 'robot' to the wall.

The students can now engage in a joint brainstorming and discussion session: How accurate were they? Why did errors happen? Did the designated robot not accurately follow their instructions? Did another group do a better job than us? Does the problem of accuracy also occur when another student assumes the role of the robot and receives the same instructions?

Once the brainstorming session is complete and students have addressed some of these concerns, you can move onto a mini-competition. For this competition, several groups of students will be asked to create instructions for the same route. Depending on the time available and the teacher's approval, the distance to be covered can be longer or shorter or more challenging, for example from one location to the classroom door, around a table or returning via a different route. Similar to the test run, the robots must follow the instructions of their team and any deviation from their goals will be recorded and compared. The winning group (with the least deviation from their goal) gets to explain their strategy to the other groups. They must outline how they minimised the deviation from their goal.

If it has not occurred to your students yet that movement should be standardised as precisely as possible, for example, selecting length of shoes as a unit instead of steps), you as a teacher should suggest this now. There is an opportunity here for you to take a little detour to history if needed: the transition from older forms of measuring units such as cubit and feet to the Parisian metre was, and continues to be, a decisive one.

Comment: When the students experience how their varying height and step length (gait) make it difficult to give precise instructions to the robot they are led to the problem of standardising lengths. Letting students pose as robots operated by remote control adds a hands-on and motivating touch to the problem. LEGO provides teaching resources to support such activities. See https://education.lego.com/en-gb for more information on this.

4.7 Presenting Balance Sheets in a Positive/Negative Light

Overview: Looking for errors or distortions in public accounts is a good opportunity for applying our knowledge of descriptive statistics. In this activity we are aiming to highlight for our students how different perspectives influence how the data may be interpreted. This may also involve us asking the question: how do we best present the data we have collected so that it will best serve our purposes?

> How would you create a problem based on this scenario?
> Please jot down some ideas so that you can compare your suggestions to ours.

Implementation: In Table 4.4 we present the simplified balance sheet of a company and inform students that the company is now up for sale. The value of the company is dependent on the profit the company has made in the past number of years. Therefore, the company will attempt to present itself in the best possible light. Potential buyers, on the other hand, would be inclined to view this information from a more critical perspective. Table 4.4 summarises the profit made by the company (in millions) over the last 10 years.

Figure 4.9 shows how a simple graph, such as a bar chart, can be used to illustrate these figures.

Table 4.4 Company profits for the last 10 years

Year	Profit in Euro (000,000's)
2008	198
2009	172
2010	145
2011	132
2012	58
2013	122
2014	145
2015	174
2016	175
2017	176

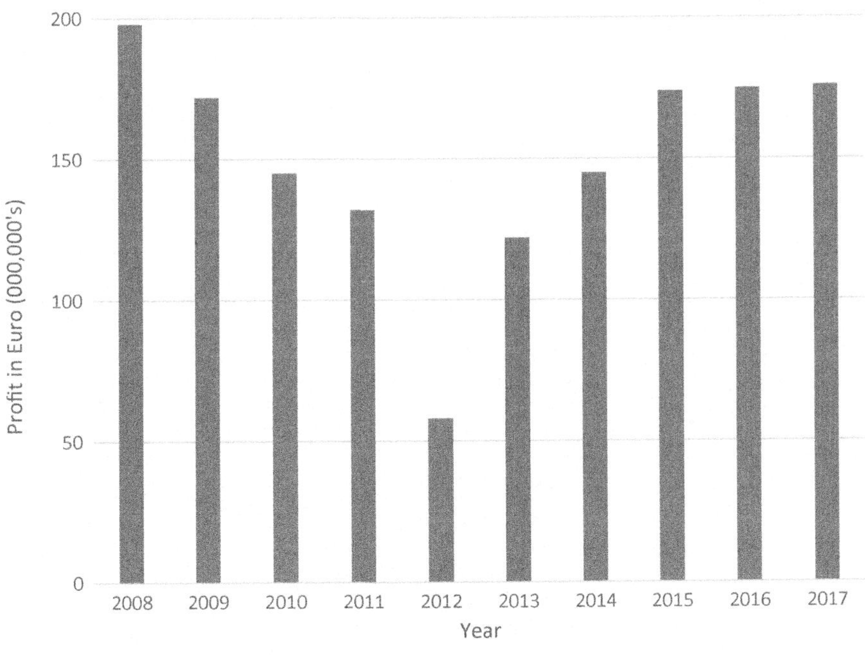

Fig. 4.9 Company profit (2008–2017)

At this stage you can split your class into two groups with one group representing the company and the other group representing the potential buyers. The group representing the company is supposed to present the figures in a positive light, while the other group should look at the data and attempt to present it in a less favourable manner. After discussing the optimum way to represent the data, each group will construct their own representations of the data and present these to the larger class group. During each presentation the other group must critically analyse the representations utilised and expose any misleading information presented.

The following figures are some examples of what could potentially be offered by the group analysing the data from the company's perspective.

In Fig. 4.10, the years prior to 2012, during which time the profit was decreasing, are conveniently omitted. A different approach is taken in Fig. 4.11.

The combined area as a whole appears more impressive than individual bars, although it is not necessarily so. When critically analysing the validity of this graph there are two additional points to consider; we calculated and displayed the total profit as a continuous variable whilst simultaneously we have opted to expand the y-axis to make the area appear even larger.

In Fig. 4.12, the cumulative profit for the company is displayed. Again the y-axis has been expanded to distort the overall size of the profit, while the cumulative profit figures themselves are misleading.

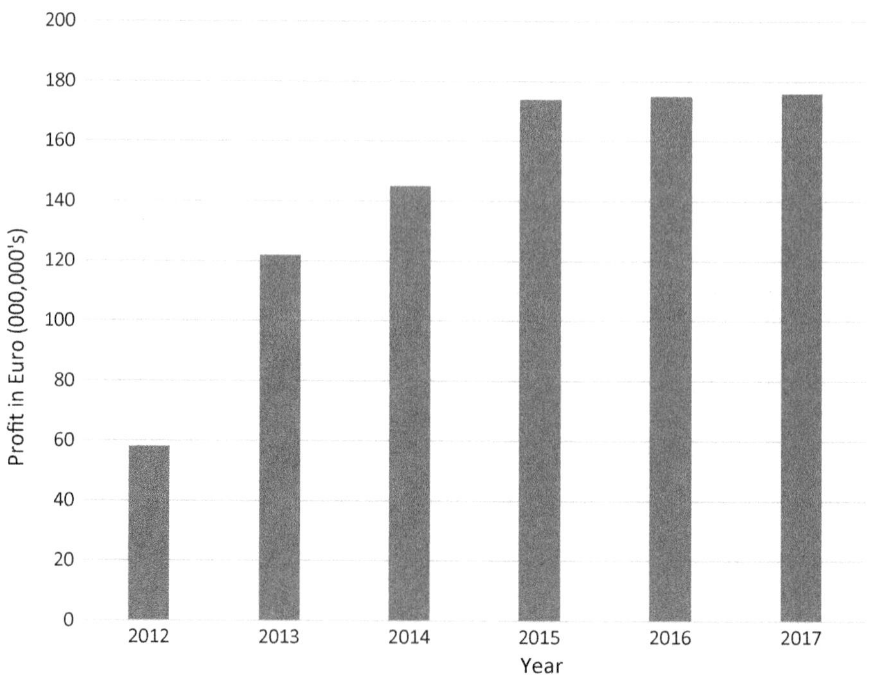

Fig. 4.10 Alternative presentation of company profits (2012–2017)

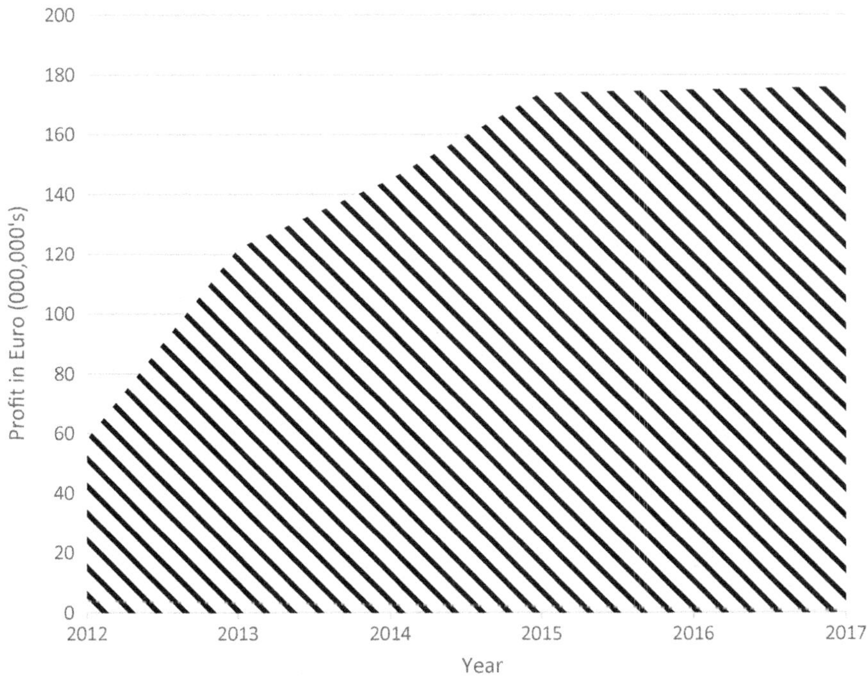

Fig. 4.11 A stacked area chart of company profits (2012–2017)

Now it is your turn.
Can you come up with more representations that display the company's profits in a positive manner?

We will now briefly look at this scenario from the viewpoint of the buyers and come up with a potential chart (Fig. 4.13) they could use to represent the data.

Figure 4.13 shows that the company profit history appears faltering at best. After the successful year of 2013, the company experiences several years of a downturn in profit.

Now it is your turn.
Come up with some other representation from the perspective of the buyers that would allow them to depict the company in a negative light.

Comment: We firmly believe that this innovative approach to data analysis motivates students to look at different ways of illustrating data. People who have

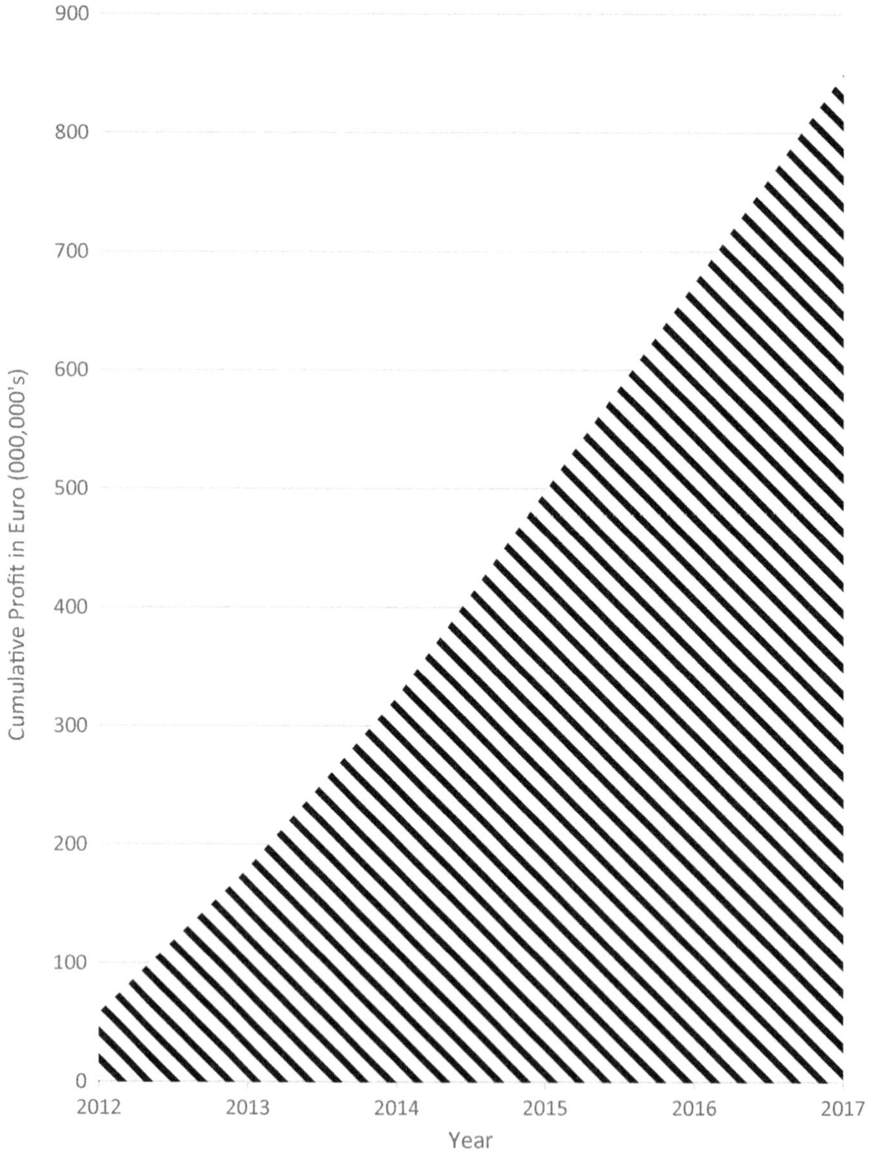

Fig. 4.12 Cumulative profit (2012–2017)

experience constructing graphs in which they intentionally mislead the reader are more likely, and better able, to identify charts that have been manipulated by others.

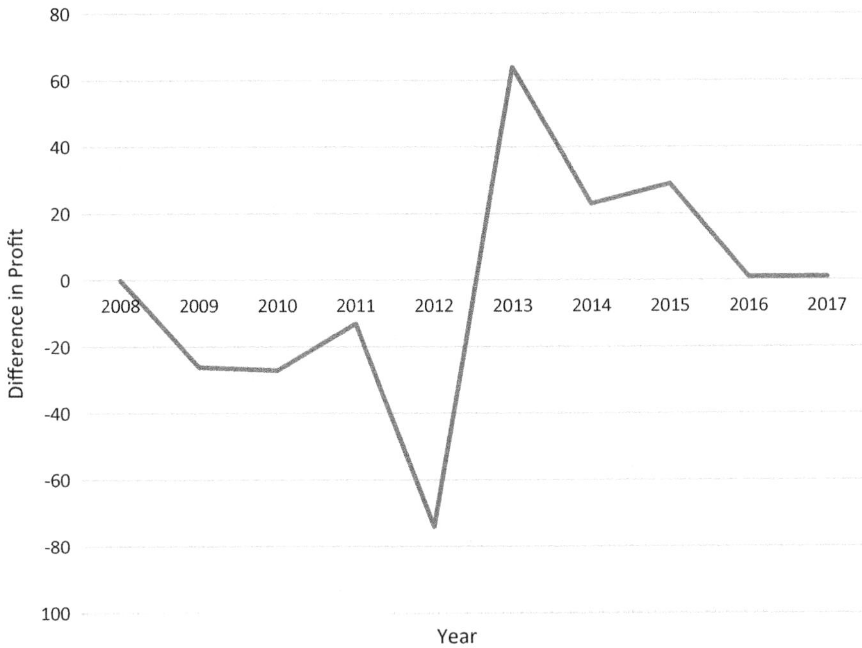

Fig. 4.13 Difference in company profit (2008–2017)

4.8 Deceptive Packaging—Estimating Volume

Comparing stated volume to actual content of a package has been investigated numerous times in mathematics education but it is still a worthwhile activity to engage in with your students.

Overview: People who are able to assess the volume of products correctly are less likely to be deceived by elaborate packaging and advertisements and can potentially save themselves a lot of money when shopping. Calculating the volume of cuboids is a necessary skill when dealing with such problems. This skill can help us assess how closely the stated volume resembles the actual volume of the product bought. Similar problems can be created using packages that are in the shape of cylinders, pyramids, or any other three dimensional shape. Bear in mind that in the case where the volume is too difficult to calculate directly, it can be measured by immersing it in water, similar to the technique first outlined by Archimedes.

How would you create a problem based on these ideas?
Please jot down some ideas so that you can compare your suggestions to ours.

Implementation: We suggest that you invite your students to bring cuboid-shaped product packaging, for example, a cereal box, with them to their mathematics class. The dimensions of the cuboids will be measured and then the volume calculated. Then we will compare our result to the volume stated on the package. What proportion/percentage of the cuboid is taken up by the actual content? Are there plausible reasons for any differences noted, such as thermal expansion? Can products with a particularly large difference between their stated and actual volume be categorised in a certain way? For example, we might expect the greatest differences to be in the case of luxury products. In other cases the packaging only serves to wrap the goods, for example flour, sugar, and butter are all packaged without any additional space around the product. In such cases the actual and stated volume are almost identical. Cosmetics and toys on the other hand are often packaged in a way suggesting more content than is actually there. In such cases it makes sense to pay attention to the small print. Typically this small print, written somewhere on the packaging, outlines exactly what the package contains.

Comment: This example of deceptive packaging is one that is easy to relate to real life. It is also possible to approach this problem from the opposite perspective. Imagine you want to sell perfume. How would you package it? If you search 'deceptive packaging' on the internet, you will find lots of websites focused on this issue:

https://venividivulgo.wordpress.com/2012/11/23/deceptive-packaging-want-to-pay-more-for-less/

4.9 Treasure Map

Overview: A typical pirate movie involves a treasure map. On these maps you will either find the location where the treasure is hidden, marked with an 'X', or instructions on how to reach this location. Obviously in order to be able to find this location, a treasure map needs to be much more detailed than just a drawing/picture.

> How would you create a problem based on this idea?
> Please jot down some ideas so that you can compare your suggestions to ours.

Implementation: Drawing an accurate and detailed treasure map can be an interesting educational task for students. After a short introduction to the task, we would suggest listing factors that contribute to a good treasure map. These factors can also act as our criterion for assessment at the end of the activity. Obviously accuracy is particularly important here. Instructions like 'on a beach in the eastern part of the island' will not suffice.

We suggest dividing the class into small groups and having each group select a spot where they will hide their treasure. Each group has to draw a map outlining how to locate their hidden treasure and then exchange it with another group. Each group must then try to find the other group's treasure. Ideally this activity should be conducted outside the classroom but if this is not possible then the treasure (some candy or simply a mark) can be hidden in the classroom. Depending on how much time is available, more than one group can evaluate each map and then assess it. In the end, the class will decide which map is the best based on the criteria they initially outlined.

Comment: This activity can be utilised as a change from more traditional lessons as it challenges students to engage with all their senses which can ensure a long-term learning effect. Should the question of accuracy need to be dealt with in more detail, reference can be made to standardising units of length (provided steps or length of feet were used as units of length) or geography (drawing maps or scales).

4.10 Painting the Classroom

The idea for this example was developed as part of a project called 'Power Girls' (See: http://www.edugroup.at/praxis/portale/powergirls/projekt.html. Please note that this website can be translated into the English language).

Overview: Our classroom is to be painted a new colour. Before doing so, the area to be painted has to be measured and calculated accurately. We have to calculate the area to be painted so that we can buy the necessary amount of paint. The level of accuracy feasible in this case also needs to be taken into account as those who happen to miscalculate the area will end up buying too much, or too little, paint.

How would you create a problem based on this scenario?
Please jot down some ideas so that you can compare your suggestions to ours.

Implementation: To begin the teacher might present the following, or a similar, scenario to the students:

Today the principal informed me that all the classrooms in the school are to be painted. I figured this would be a good opportunity for us to contribute to the decision-making in relation to which colour(s) to use. As the school only receives a small budget for maintenance and renovations, we cannot afford to waste any paint. Therefore, we need to accurately determine in advance how much paint we are going to need.

We need to be mindful of two things when we are deciding on the most effective approach to solving this problem:

1. The room has to be accurately measured.
2. How much paint per square metre is needed?

The classroom is to be measured by all students (in small groups would be best). In order to do this effectively, the surfaces that have to be painted needs to be determined, the dimensions of these surfaces need to be identified and we need to decide on how best to measure these dimensions. When determining the surfaces to be painted we may ask ourselves whether the floor, wood panelling, or radiators are to be painted. Should the ceiling be repainted? The answers to these questions will then allow us to determine what dimensions we actually need to measure?

Obviously we need to measure the width, length and height of the room. The net area to be painted can then be determined by subtracting from the total area of the room the sub-areas that do not need to be painted such as windows, doors, closets (depending if they are affixed to the wall or not), and so on. How about small sub-areas such as light switches or power sockets? First measure them and then include them in your discussions on the net area to be painted.

Remarks: The teacher can adjust the level of difficulty by making the conditions simpler or more complicated for example, whether to include small sub-areas or not.

Once the parameters of the problem have been determined, students set about measuring the walls and the ceiling (by measuring the floor) in order to calculate the total area. Please have your students record all measurements and calculations so that it is easy to compare them in the end.

At this stage the teacher moves from group to group, monitoring cooperation within the groups (everyone should be involved) and providing assistance where needed.

Presenting the results will be the final step. Each group will be asked to present their findings regarding the area of one particular part of the room for example, walls, ceiling or sub-areas. Every group will compare the results presented to their own results and confirm, or challenge, the accuracy of the result. If doubts remain, the measurements will be re-calculated, by the entire class group, in order to reach a conclusion that everyone agrees upon. The class will now give careful consideration to the final result and the approach used to determine it. They should consider the following questions: How accurate is our final result? How significant would the overall error be if all the lengths were measured one millimetre too short or too long? What impact does the error have in relation to the problem?

This would be a good time to introduce some information regarding the purchase of paint. Paint is not sold per millilitre. In general the smallest quantity sold is 2.5 l—this will typically cover 14.25 m^2. To verify these values we would recommend a visit to a hardware store, or an internet search (the following website offers information regarding coverage rates of paint http://www.duluxprotectivecoatings.com.au/technotespdf/5.3.1%20Volume%20Solids%20and%20Spreading%20Rate.pdf). At this stage students may protest and ask why we needed to measure our dimensions with such accuracy. A possible answer is so that they will learn to consider the effects of accuracy and errors when working with a hands-on problem such as this.

Comment: The impact of the result is the key issue when dealing with such projects. If a calculator is used and an error is made that results in 30 tins of paint being purchased instead of 3, the project might become very expensive if nobody notices the mistake in advance of the purchase. If a mistake like this occurred when solving a typical textbook problem, the worst that will happen is that they will be awarded no marks. The impact of this is far less than being stuck with 27 extra tins of paint that you do not need. This is a great opportunity for students to realise that inaccurately applying mathematics in real life can have consequences, and someone will have to account for such errors. For example, in a case like this where too much money is spent, the question arises regarding who is going to cover the additional cost.

4.11 Golden Ratio

Teachers have possibly spent many hours thinking about the Golden Ratio and how to best introduce it in their mathematics classroom (see some suggestions at https://www.theproblemsite.com/reference/mathematics/the-golden-ratio/ or http://www.totemlearning.com/totemblog/2015/4/22/the-golden-ratio-a-brief-introduction).

Overview: We suggest a mini-competition here to show some ways for teachers to engage with the Golden Ratio as described in mathematics education literature.

> How would you create a problem based around the Golden Ratio?
> Please jot down some ideas so that you can compare your suggestions to ours.

Implementation: To begin we assume that the Golden Ratio has already been covered in class and so we move on to a mini-competition. Split your class up into small groups of no more than 3 students. The aim of the competition is to identify as many examples of the Golden Ratio as possible in a photo of a famous or local landmark, for example the Taj Mahal (https://upload.wikimedia.org/wikipedia/commons/c/cb/Front_View_of_Taj_Mahal.JPG). Each example identified by the groups has to be clearly marked on the photo and all ratios and calculations clearly shown!

After a trial run is implemented to demonstrate the process using a famous or local landmark, the teacher and students are ready to begin the competition. Each group is instructed to select a random photo or picture and find examples of the Golden Ratio in it before passing it on to the next group. The next group scores a point if they find at least as many examples of the Golden Ratio as the previous group from whom they received the picture or photo. Depending on how much time is available, several rounds can be played and the cumulative score can be calculated.

Comment: Competitions such as this could also be used for cross-curricular purposes—for example, a similar competition could be run in cooperation with an art teacher by looking for the Golden Ratio in the works of famous artists, for example: https://www.goldennumber.net/art-composition-design/.

4.12 The Sundial

When asked for the time you will probably look at your watch, your phone or the nearest wall clock to provide an answer. Have you ever considered the knowledge and technology needed to inform you about the actual time? To begin investigating this issue we will first need some basic definitions about time, measuring time and norms. For example, what is a year, a day, an hour? Furthermore, have you ever considered how people thought about time in the past? Our colleagues in physics worked hard to define the concept of a second by looking at radioactive decay (see: https://en.wikipedia.org/wiki/Second). In addition to this, the elder definition of a year was done by astronomy because the planet Earth needs one year to move around the sun. Looking at the development of instruments to measure time is a very interesting endeavour. Historically, knowing the exact time was very important for travelling, especially when sailing for long periods of time on the ocean. On the other hand, the needs of shipping were an important motivation to improve clocks and other instruments that told the time. As such, navigation is a very nice theme for reality-based mathematics education.

Overview: We ask you to imagine living in the past, maybe a few hundred years ago. Think about a nice summer's day in a little town, and someone asking you *what time is it now?* You want to give an accurate answer, not an estimation. What do you do? There is no watch on the wall or on your wrist—things like these will only be invented in the future. A computer screen, a mobile phone, the internet and other electronic devices are objects that also do not yet exist. Maybe you know that there is a clock on the church tower in the centre of the town. If you are a lucky and wealthy person you may be able to see this clock everyday by looking out of the window of your house. Alternatively, if this is not the case you could direct someone to the centre of town where they will be able to view the church tower themselves.

Now we present another way of answering the question posed above. Have you ever seen the object presented in Fig. 4.14?

This is a "folding sundial" invented by Georg von Peuerbach (astronomer in the service of the imperial court of Friedrichs III) in 1451. For about 400 years this was the pocket watch for rich people. But how did it work? Please try to find out for yourself!

We remember that the main feature of a sundial is 'something' that casts a shadow on a scale, from which we can read the actual time. In our opinion the string in the folding sundial has two functions. The first is to stabilise the object (i.e. it helps to connect the two parts) and the second is to cast a shadow. When the sun is shining the shadow of the string will fall somewhere on the scale on the lower part of the sundial shown in Fig. 4.14. We can see the number 12 near the hinge that connects the two parts of the sundial. On the right side of the sundial, we see the numbers one, two, three and so on. This is what we would expect if looking at a modern watch. You might also be wondering what is the round object in the middle of the lower part? This is a compass. The compass seems to be important but do you know why?

Fig. 4.14 Folding sundial (https://www.steiermark.com/de/steiermark/ausflugsziele/museum-fuer-geschichte_p7789)

We did not find any instructions informing us how to operate the sundial, but we were able to devise our own, as outlined below:

If you want to know what time it is:

1. Open the folding sundial;
2. Using the compass, find out in which direction is North;
3. Turn the folding sundial in this direction;
4. Look at the shadow of the string and read the time.

We know that you need sunshine to do this. We have no doubt that this is one important reason why people today use watches or other devices instead of folding sundials.

When using a sundial the user is not required to understand how and why it works. This is like the situation today. If, for example, you read 9:50 on a screen of an electronic device you know that in 10 min time it is 10 o'clock. You do not need to understand anything about the technical aspects of displaying the digits on a screen. However, developing an understanding of how the sundial works presents us with some ideas for mathematical tasks suitable for the classroom.

Implementation: If you would like to understand more about sundials and construct your own sundial, you must change your perspective and be willing to investigate the mechanics of the sundial. Ronald Hohl has recently completed his thesis at the University of Linz about sundials, entitled Sol 365—Project-oriented construction

Fig. 4.15 Basic sundial (Hohl 2016, p. 73)

of a sundial for secondary school level (http://epub.jku.at/obvulihs/content/titleinfo/1960988). In it he has constructed the very simple and cheap sundial shown in Fig. 4.15.

To replicate this folding sundial, you (or your students) need a cardboard carton/box and some string. However, you may notice some key elements of the sundial are missing in Fig. 4.15. If we compare our simple model (Fig. 4.15) to the folding sundial invented by Peuerbach (Fig. 4.14) we notice that we are missing a scale representing the hours and a method to find out what direction North is. Hohl (2016) took some simple steps to overcome these design issues, as shown in Fig. 4.16.

To upgrade his sundial Hohl (2016) used a compass to orientate his sundial to the north. He then took a piece of paper and used his own digital watch to draw lines where the shadow of the string was cast for each hour on the paper. We think that your students may adopt a similar approach—maybe with a little help and guidance from you. They can use a smart phone and GPS instead of the compass if they so wish.

Now your students (and you) must decide if you are happy with these results (you have built your own sundial that shows the time). It is not as exact as our electronic watch but it works nonetheless. If you would like to know more and to build a better version of a sundial you will need to learn some additional mathematics and carry out some extra work.

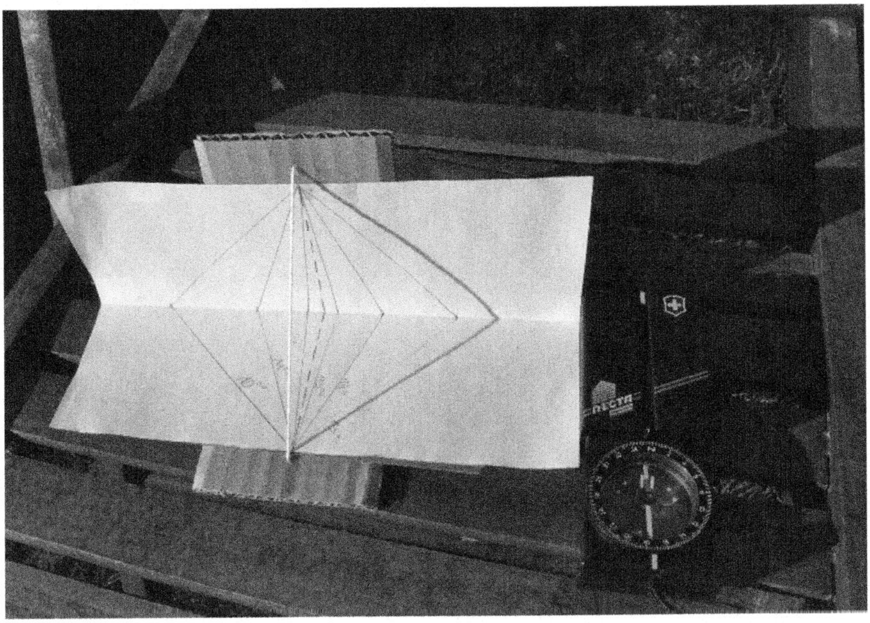

Fig. 4.16 Upgraded sundial model (Hohl 2016, p. 75)

From our point of view the main question that now arises is: where is North? A compass shows the direction to the north magnetic pole, however a sundial needs the orientation to the north geographical pole. The difference between these directions depends on your location on Earth. The north magnetic pole is located somewhere north of Canada but it is not the same location as the north geographical pole as shown in Fig. 4.17.

In 1994/1995 a person living in Cambridge Bay in Northern Canada would have found the same direction to the north magnetic pole and the north geographical pole. A person living in Europe during this time would have had to be more careful. From their perspective there would have been a significant difference between the location of the poles. From Fig. 4.17 we see that someone living in the town of Alert (the northernmost, permanently inhabited place in the world) should not use a compass since the direction of both poles differ significantly. We will not explain this here but for a mathematics teacher the geometrical aspects of the question are clear: What is the angle of the direction to the north magnetic pole and the north geographical pole depending on your position in the northern hemisphere? If you would like to engage with this question as a two-dimensional problem, you will need maps (projections of the surface of the Earth). On the other hand, if you would like to employ more sophisticated approaches, you may use a three-dimensional geometrical approach.

One reason to conclude this modelling project at this point is that we are not the first to examine the links between geography and mathematics to gain a deeper understanding of this area. Instead, we direct you to some websites which may act

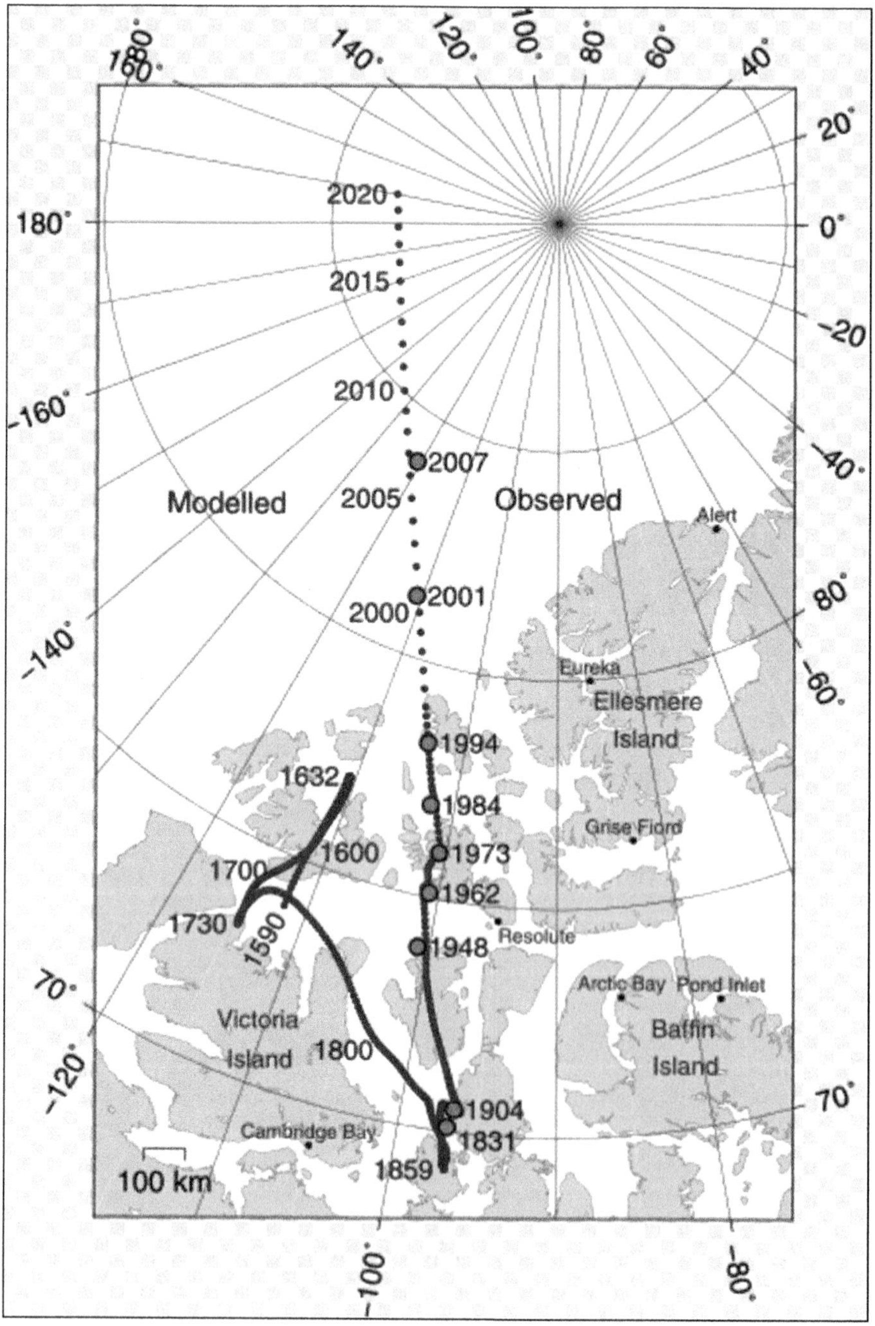

Fig. 4.17 Changing location of magnetic north (https://en.wikipedia.org/wiki/North_Magnetic_Pole)

as a valuable resource and provide you with a series of hints for incorporating this type of project into your mathematics class.

https://www.sunearthtools.com/dp/tools/pos_sun.php?lang=en

https://www.geogebra.org/m/NNMPQaag

http://news.nationalgeographic.com/news/2010/12/photogalleries/101228-sun-end-year-analemmas-solstice-eclipse-pictures/

http://www.geoastro.de/GeoAstro/GeoAstro.htm

http://sundials.org/index.php/teachers-corner/sundial-mathematics

http://www.bowlandmaths.org.uk/projects/sundials.html

4.13 Conclusion

In this chapter we have presented twelve potential reality-based mathematics activities derived from everyday occurrences. This chapter built on the approach outlined in Chap. 3 and sought to demonstrate how simple, everyday events can be utilised by teachers to develop meaningful activities for their class group. In the next chapter we will look at extending this idea to develop a more substantial modelling task which can be implemented over a longer period of time. However, if you would like more ideas similar to those presented in this chapter we recommend the following book: *Real-World Problems for Secondary School Mathematics Students: Case Studies*, which was edited by two of the authors of this book (Maasz and O'Donoghue 2011). In this book you will find a range of ideas for modelling tasks taken from a variety of contexts such as space travel, sports and the sciences.

References

Androsch, G., Aspetzberger, K., Boxhofer, E., Lischka, U., Lindner, A., Maaß, J., et al. (2015). Maths eyes: Die Welt mit mathematischen Augen sehen. Bericht über einen Wettbewerb als Anregung für motivierenden Mathematikunterricht, In: L. Del Chicca, M. Hohenwarter (Eds.), *Mathematikdidaktik im Dialog*. Linz: Trauner Verlag.

Brandl, M., Brinkmann, A., Bürker, M., Maaß, J. (2013). Mathe vernetz: Anregungen und Materialien für einen vernetzenden Mathematikunterricht. Band 3 (Deutsch). Munchen: Aulis Verlag.

Hohl, R. (2016). *Sol 365—Project-oriented construction of a sundial for secondary school level (doctoral dissertation)*. Linz, Austria: University of Linz.

Maaß, J., & Wildt, M. (2012). Networking is worthwhile: Sustainable mathematics teaching helps to save time! (Vernetzen lohnt sich: Nachhaltiger Lernen hilft Zeit sparen!) (German). In M. Brandl, et al. (Eds.), *Mathe vernetzt. Anregungen und Materialien für einen vernetzten Mathematikunterricht. Band 2* (pp. 9–22). Köln: Aulis Verlag.

Maasz, J., & O'Donoghue, J. (Eds.). (2011). *Real-world problems for secondary school mathematics students: Case studies*. Rotterdam: Sense Publishers.

Maguire, T., Meehan, C., O'Mahony, J., Ryan, M., & O'Donnell, C. (n.d.). Developing maths eyes—A resource pack for teachers. Retrieved November 21, 2017, from http://www.haveyougotmathseyes.com/wp-content/uploads/resources/mathseyes_resource_pack.pdf.

Chapter 5
An Example of a More Extensive Project

In the past two chapters we have attempted to show how modelling processes, based on textbook problems, can be initiated. We also demonstrated how a mathematician's view can help us identify problems suited to mathematics education, which students can easily relate to. All this enables us, as teachers, to develop modelling activities for classrooms and allows us to involve students in the learning of mathematical content and to develop their mathematical competencies. We are aware that using real situations, issues and challenges for school-based mathematics teaching and learning draws us into the debate on the selection of suitable themes for school education. We disagree with the position that school students should not be exposed to real issues and concerns in their education and with the position that education is a panacea for all humanity's ills. We think (and we know from our teaching experience) that mathematics can help students and teachers to better understand a lot of important aspects of our real world. But understanding is not doing or acting. For example, if students learn about the risks of gambling, smoking, drinking, consuming drugs, and driving fast by doing real world mathematics projects they will know better what kind of behaviour or actions might be appropriate and what might be dangerous, but this is not a guarantee that they will behave or act in a particular way.

Those who are equipped with a problem solving/modelling perspective on teaching mathematics will not only encounter problems that can be put to use without much effort, but will also come across topics they are not yet familiar with, topics that do not appear to have much mathematical content or topics that at first seem too complex to be suitable for the classroom. This can act as a deterrent. From our experience however, we know that teachers from German-speaking countries willing to use such problems in their classrooms often get support from scientists and experts on the didactics of mathematics. These experts conduct similar projects themselves or they support and supervise activities such as modelling days, a modelling week or individual modelling projects by contributing their ideas.

Our extensive experience tells us that after a few supervised attempts these teachers no longer need our assistance in providing motivating and successful mathematics education. The teachers who have been cooperating with us long-term often come up with their own ideas and they are also able to assist other colleagues in this endeavour. In many cases the scope of the projects put into practice in mathematics

J. Maaß et al., *Mathematical Modelling for Teachers*, Springer Texts in Education, https://doi.org/10.1007/978-3-030-00431-6_5

classes grew together with the experience these teachers gained in modelling and applied mathematics education.

In this chapter we will present a project idea to illustrate how modelling projects can be implemented over a longer period of time. We have selected gambling as the theme for this project. Most people become aware of gambling through advertisements which state the odds or quotas for a certain event (usually sporting events) and therefore in order to understand gambling you first need to understand odds or quotas and the mathematical underpinnings of these concepts. This is addressed in the early stages of this chapter using two examples based on real life data. Once these mathematical ideas have been introduced and understood we then use modelling to explore a variety of different sports betting scenarios.

More extensive projects like this do not necessarily have to be the sole focus in every class over several weeks. Instead, they can be split into sub-topics, which 're-emerge' from time to time, like a spiral, and can then be expanded and integrated with new aspects. Since such projects focus on student involvement, we would like to point out that this cannot be conducted in an uncoordinated manner. In fact, quite the contrary, as putting such projects into practice requires thorough planning and structure. Methods for implementing such projects have been discussed extensively in previous publications or can be found online. For general information on this topic we recommend the following website which offers resources and ideas (https://maths.org/) or a decree by the Austrian Federal Ministry of Education (https://www.bmb.gv.at/ministerium/rs/2001_44.html) which can be translated into English on the website. A synopsis of the key stages of a modelling project on this website are outlined in the next section.

We are aware that implementing teaching activities such as extended projects present challenges for teachers and schools. But we also know that projects like this have been done many times with overwhelming success. Teachers tell us about positive change in the general motivation for mathematics, and students tell us that they decided to study mathematics or sciences or technology because they enjoyed learning and doing mathematics in this new way.

We know that teaching conditions are really different in different countries and even in different schools within the same country. In some situations, it is impossible to start bigger projects for a variety of reasons related to local and national circumstances. In other situations it seems to be impossible because of competing demands on the curriculum that lead to reduced time for teaching mathematics. But our considered view and experience tell us that it is worth the effort. Therefore, we introduce a longer project here to provide such an experience for teachers and students, and build in possibilities for breaks or stops that can make the experience more manageable. You, or better still the working group, can decide to stop or to search for more data or interesting extensions at another time independently or together as a class group.

5.1 Project Stages

We propose that when commencing any large project that you approach it by following the stages listed below. We shall also illustrate these stages using an example in the coming pages.

Stage 1: Project Idea/Topic Creation
It is important that the interest of all participants be considered and that there is sufficient time allocated for teachers and students to agree on the topic they intend to work on, or the problem that they want to solve.

Stage 2: Defining a Goal and Planning
In defining the goals, various interests will become apparent, sub-topics may be discussed and the goal to be attained will be agreed upon. The scope of the project and the available resources must be analysed and taken into consideration when planning. Key roles/tasks must also be identified at this stage and different individuals/groups assigned to these roles.

Stage 3: Preparatory Stage
This stage involves the gathering of all relevant information as well as any materials/resources that are required. This time may also be used to plan excursions, hold discussions with experts, and do further online research such as watching videos. Changes to our proposed project plan may become necessary at this preparatory stage.

Stage 4: Project Implementation
At this stage, it is necessary to find out as much as possible about the topic at hand. The plan derived in previous stages will now be executed by the students in various group settings with minimal teacher instruction. Teachers can provide assistance as consultants, experts, and conflict managers. During this period, it is particularly important for students to exchange experiences and interim results and to discuss the problems encountered and what could be done to better coordinate the project. It might also become necessary to check the progress of the project and the emotional state of the students.

Stage 5: Project Presentation/Documentation
Modelling activities require a formal closure of the project. Everyone involved in the project should get the opportunity to present his/her results to their peers and teacher, and to a larger audience, if possible. It is important at this stage of the process to provide constructive feedback to the groups so as to ensure project results are fit for presentation. Documentation is also an integral part of bringing a project to completion; it is the basis for presentation, communication, contemplation, and evaluation. Therefore, it should contain all the important results, project stages and the experiences of people working on the project.

Stage 6: Project Evaluation
The final evaluation serves to monitor the project results and to develop the quality of future projects. The following questions helped us define our goals during the

planning stage: What do we want to attain? To what end? Using which tools? The data collected now allows us to analyse the answers to these questions in a systematic manner both during, and at the end of our project.

During this final stage, the experiences of the participants and the ongoing processes will be discussed. This reflection is an integral part of the project evaluation and fundamentally, it should be done by the people involved in the project. However, in order to avoid 'blind spots' during some evaluation stages, it is also necessary to include an outsider's view.

There is a lot of pedagogical research done about this method of teaching. We will not attempt to write an overview here. We think it is more helpful to show an example. This example demonstrates how to initiate learning that is activity-focused and promotes a deep and connected understanding of mathematics that contributes to students' long term learning.

5.2 Project: Sports Betting from a Mathematical Perspective[1]

Once again we enlist your active participation:
What do you know about sports betting?
Do you participate in sports betting yourself?
Do you have any idea how the odds (sometimes 'the odds' are given by 'quotas') are determined and what they imply?
Do you have a positive or negative attitude towards sports betting?
How would you feel if you found out that some of your students gamble for money, say on the internet?
Please jot down a few thoughts.

What have you written down so far? Have you done some research on the web and seen how billions of euros were generated by sports betting companies in the past year? Have you come across any current newspaper reports on people addicted to gambling? What about match-fixing scandals? Have you noticed that the odds offered for the front-runners are much smaller than those for the outsiders?

As usual we start by providing a motivational element in our proposal for using the topic of sports betting. Major sporting events like the Olympics or the FIFA World Cup receive huge media attention, and they also attract a lot of attention in schools. They can be used to deal with particular aspects of a mathematical topic and to familiarise students with them. Mathematics teachers often feel left out at this point, because such major events do not appear to provide enough material to be

[1] Jürgen Maaß and Hans-Stefan Siller have published several papers on this subject. This is a globalised version of the basic ideas and includes some new aspects.

used in class so that students can gain further mathematical insights. Frequently, they seem of to be useless from a mathematical standpoint or appear to be too challenging. However, this is not always the case! At this point, we would like to encourage you by asking you to engage in a short exercise.

Please consider if any major events have occurred recently that could be incorporated into your mathematics class.

Major events, other than sporting events, lend themselves naturally to exploitation from a mathematical viewpoint. This is evident when we attempt to explain their background and their connections to mathematics. Just think of conventional classroom exercises on half-life periods when introducing exponential functions, or the tsunami that led to the catastrophe of Fukushima in 2011. Please also consider other examples such as weather forecasting and the climate; the environment or future energy supplies; reports on the economic situation; the development of new technologies and electronic entertainment; nutrition; and fashion. But let us return to sport! There are plenty of mathematical aspects involved here as sport is about motion, scores and tables, success statistics (in tennis they are shown after each set) or training methods and all these depend heavily on mathematics.

We have selected an example from soccer for the following activity which involves a lot of money and we enter a caution here about the danger of a serious, unwanted personal/social consequence, betting addiction. The FIFA World Cup, the UEFA European Championship, the UEFA Champions League, and National Football Leagues, such as the English Premier League, are all recurring events and involve a lot of money. Wherever that is the case, mathematics can usually provide a better understanding of the cash flow involved. Therefore, we decided to consider sports betting from a mathematical perspective as our larger project and it receives attention in this chapter.

Why this topic? There are several motives for doing this. We will begin with a pedagogical one: dealing with this matter from a rational perspective in mathematics class hopefully counteracts a tendency towards irrational gambling or the imminent loss of money or happiness caused by compulsive gambling. There is also the view that sports betting is easy to understand from a mathematical perspective, provided only simple bets are considered. We shall not consider the idea of combined bets (accumulators) although they could, with a little bit more advanced mathematics, definitely be analysed if you so desired.

In this book we have chosen sports betting as an example of a more extensive project comprised of smaller sub-projects that can be dealt with independently. As opposed to other topics such as energy and the environment, nutrition or traffic control, it does not require a lot of background knowledge.

5.3 How Do We Get Started? (Project Start)

Before commencing this section it is worth briefly talking to you about the different formats that are used for expressing odds:

Decimal odds: This is the format that is typically used across mainland Europe and is fairly easy to understand. Decimal odds are simply expressed as a number in decimal form, for example 1.50. This number is the ratio of the pay-out to the original amount that you placed on the bet. For instance, with odds of 1.50, if you bet €40 (this is known as your stake) and win, you will receive €60 (pay-out) which is 1.50 times your original stake. In this case, you will make a small profit of €20 on your original €40 stake.

Fractional odds: This is the format generally used in the United Kingdom and Ireland. In this case the odds are expressed as a fraction, like 7/2. This represents the ratio of the profit (not the total pay-out) from a successful bet to the stake. Essentially this means that you will win €7 for every €2 gambled. For instance, if you bet €100 on something with 7/2 fractional odds and win, your profit will be 7/2 of your original stake and the pay-out will equal the profit plus the original stake. In this case, your profit would be €350 with a total pay-out of €450.

Moneyline odds: This is the system that is favoured in the Unites States. Moneyline odds are expressed as a number preceded by a minus sign or a plus sign, like −60 or +20. A number preceded by a *minus sign* indicates how much you need to stake to make a profit of $100. On the other hand, a number preceded by a *plus sign* indicates how much you will win if you bet $100. For instance, if you wager $50 with odds of −200, when you win, you will get a pay-out of $75 for a total profit of $25 (remember that you needed to bet $200 to make a profit of $100 but since you only wagered $50, which is a quarter of $200, your profit is only a quarter of $100). If you wager $50 with moneyline odds of +200, you will get a pay-out of $150 for a total profit of $100, since you are told your profit would be $200 for every $100 staked. Note that your original stake is returned as part of the pay-out.

A good place to start when commencing a project around sports betting is to analyse a betting slip. The numbers on the slip, and the way they are arranged, would suggest that there is plenty of mathematics involved in betting. Our informal inquiries among school and university students though have shown that the opposite is sometimes true. Initially, school and university students do not necessarily associate mathematics with sports betting.

We have refrained from providing a screenshot of a betting agency's website here for copyright reasons and because they show real-time betting odds, but you can easily find some on the web (https://www.tipico.de/ or www.ladbrokes.com or http://www.paddypower.com/). During the soccer season bets relating to soccer matches take a prominent position on these websites. Matches about to take place are typically listed by displaying the date and time of the match and also the names of both teams involved, for example Bayern Munich versus Hamburg SV or Manchester United versus Liverpool. Next to it you can find the current odds and perhaps even some buttons offering additional information and further betting options ('Who will score the first goal?' or 'Will both teams score a goal?').

At this point it might be useful to use an example to allow us to interpret the bet presented. In the penultimate game in Group C of the 2018 World Cup Qualifying campaign, Northern Ireland welcomed Germany to Windsor Park in a game that was critical in deciding who would win the group, and in turn qualify for the FIFA World Cup (2018). The odds offered by one bookmaker for this event were:

	Home	Draw	Away
Northern Ireland versus Germany	10/1	9/2	1/4

In this case, Northern Ireland is the first team listed and so we know they are the home team. As these odds are taken from a UK betting agency website, they are presented in fractional format, although if you wish to convert to decimal format we will show you how to do this shortly. Germany are favourites to win this match at odds of 1/4 (pronounced 1 to 4 or 4 to 1 on), which means that if you were to wager €1 on Germany to win and they did win, then you would receive €0.25 in profit for a total pay-out of €1.25. The odds of the match ending in a draw are given as 9/2 so if you staked €1 on this outcome you would make a profit of €4.50 if it actually happened, with a total pay-out of €5.50. Finally, the bookmaker offered odds of 10/1 for a Northern Ireland win. In this example for every €1 wagered on Northern Ireland to win the game the betting agency is offering a profit of €10 leading to a pay-out of €11.

Apparently, the betting agency views Germany as the front-runner in this match, whereas Northern Ireland is seen as the 'outsider'. The higher the pay-out per euro wagered the less likely a team is to win, as seen from the perspective of the betting agency.

If we express this interpretation in mathematical terms, it means that the *implied probability* that Germany wins this match is *high* according to the betting agency, whereas the implied probability of a draw is *low*, and the implied probability of Northern Ireland winning is *lower* again. At this point we have used a number of terms belonging to the theory of probability, which have already been introduced in primary school. But how can we use the given odds to calculate the implied probability? When dealing with fractional odds the process is as follows: the implied probability is the ratio of amount staked to total pay-out for a successful bet and is found by dividing the amount staked by the total pay-out if the bet is successful. In mathematics terms this idea yields the following equation:

$$Implied\ Probability = \frac{denominator}{(denominator + numerator)}$$

In our example, the implied probability of a draw is therefore:

$$Implied\ Probability = \frac{2}{(9+2)} = \frac{2}{11} = 0.1818$$

Before going any further it might be useful to show you how to convert from fractional odds to decimals odds, if you have not figured it out already. To do this all we need to do is convert the fractional odds into a decimal and then add 1, since the decimal odds tell us the total pay-out we will receive if we win, including our stake. For instance, in the example that we are looking at here, the decimal odds of Germany winning the World Cup qualifier is $\frac{1}{4} + 1 = 1.25$. To convert decimal odds into an implied probability we again divide the amount staked (always 1 in this case) by the total pay-out if the bet is successful.

$$Implied\ Probability = \frac{1}{decimal\ odds}$$

This formula is written and interpreted in mathematical terms as

$$\text{P(Event occurring)} = \frac{1}{decimal\ odds}$$

This formula, as well as the formula for converting from the other odds formats into implied probabilities, can be obtained on the web (https://www.bettingexpert.c om/how-to/convert-odds).

Now that we understand how to convert between fractional and decimal odds we will use decimal odds only going forward. However, you may of course use fractional odds with your class group during this project if you so wish. We will now present another example. Below are the odds for a 2017/18 Champions League, Group H game between Tottenham Hotspurs of England and Real Madrid of Spain.

	Home	Draw	Away
Real Madrid versus Tottenham Hotspurs	1.36	5	7.5
Odds expressed as implied probability (P):	0.7353	0.2	0.133

According to the betting agency:

- P(Real Madrid winning the match) $= \frac{1}{1.36} = 0.7353$;
- P(Tottenham Hotspurs winning the match) $= \frac{1}{7.5} = 0.133$.

Before asking ourselves (or questioning the betting agency) how these probabilities were determined, let us first perform a plausibility check. After calculating the probability for each possible outcome of the match, it might make sense to add them all!

$$\frac{1}{1.36} + \frac{1}{5} + \frac{1}{7.5} = 1.0683$$

Did we miscalculate here? The underlying mathematical theory tells us that the total sum of the probabilities associated with each possible outcome of an event has to be 1, doesn't it? Let us take a look at the same game with a different betting agency. This agency provides the following odds:

	Home	Draw	Away
Real Madrid versus Tottenham Hotspurs	1.4	5	7
Odds expressed as implied probability (P):	0.7143	0.2	0.1429

It seems remarkable that different betting agencies offer bets on the same match at slightly different odds. If we calculate the total sum of these new probabilities, it again does not amount to 1:

$$\frac{1}{1.4} + \frac{1}{5} + \frac{1}{7} = 1.0572.$$

Even if we recalculated the total probability based on the odds offered by several betting agencies, the result would remain the same—the odds have no bearing on the outcome. Therefore, there has to be another reason behind this result! We would like to present you with another task to give you a little time to think about this.

Select a match of your choice and compare the odds offered for this match with different betting agencies.
Compute the total probability for the game you selected.
What did you notice?
Describe your findings and form a hypothesis!

No matter which match or betting agency we look at, the result is always similar: the total of the individual probabilities will always be larger than 1. The reason for this is fairly simple: the remaining difference is the betting agency's profit expressed as a percentage, which it gains from each bet that has been placed. This means that for our match, Real Madrid versus Tottenham Hotspurs, the first betting agency that we looked at gains 6.83% of the total bets placed while the second betting agency only gains 5.72%.

We have already noticed a very important fact. It is impossible for us to compare the betting agencies (concerning their profit) to one another, or to determine which of the individual odds offered yields the largest proportion of the total profit! It is difficult for betting agencies, particularly those active on the web, to decide between offering better odds that will appeal to customers or odds that will result in a higher commission for themselves, but may not entice the customer to bet. This issue has become particularly important in recent years as it is now extremely easy to instantly compare the odds offered by different betting agencies online. The issues raised here lead us to question how a betting agency actually makes money.

Let us now try to summarise the topic of sports betting and perform some calculations. People bet money on a certain outcome they expect to, or hope will occur. A bookmaker accepts the money and guarantees to pay profits for the odds the money has been accepted at. In doing so, the bookmaker hopes to make a profit—if possible not just by mere luck, but by means of a percentage of turnover, known as a commission. Later on we will be looking at some cases in which the bookmaker participates in betting to counteract the risks he is taking.

5.4 Our First Modelling Task[2]

Let's begin with a simple scenario whereby the bookmaker does not seek to make a profit and instead the total money placed on an event, such as a match will be returned to the bettors who placed money on the correct outcome (this would be the simplest case; in reality such a bookmaker would be called a 'totalisator'). Assume €10,000 is placed on Team A to win, €20,000 on Team B to win, and €30,000 on a draw. Recall that earlier in the chapter we stated that the quotas were the ratio of the original amount that was placed on each outcome to the total pay-out. For a bookmaker who wishes to only break even, the total pay-out should equal the total amount staked on all the outcomes. Let's see how we calculate the quotas in this example:

Outcome	Quota calculation	Quota
Team A win	$\frac{10{,}000}{10{,}000+20{,}000+30{,}000} = \frac{1}{6}$	6.00
Draw	$\frac{30{,}000}{10{,}000+20{,}000+30{,}000} = \frac{3}{6}$	2.00
Team B win	$\frac{20{,}000}{10{,}000+20{,}000+30{,}000} = \frac{2}{6}$	3.00

The bookmaker's website or app would present this as follows (if set to decimal format):

Team A Win : 6.00 Draw : 2.00 Team B Win : 3.00

[2]From this point forward the authors will refer to "decimal odds" as "quotas".

What happens if Team A wins? For each euro placed on Team A, the bookmaker pays out €6, therefore, a total of €60,000 is paid out.

What happens if Team B wins? For each euro placed on Team B, the bookmaker pays out €3, therefore a total of €60,000 is paid out.

What happens if the match ends in a draw? For each euro placed on a draw, the bookmaker pays out €2, therefore a total of €60,000 is paid out.

In summary, whoever wins their bet receives a share of the profits according to the quota. However, the bookmaker makes no money in this scenario so it is reasonable to ask why someone would take on the role of bookmaker if he/she makes no money. After all, they have expenses for the premises, their employees, the internet and other legitimate expenses associated with running a business. Since they are in business they would also like to make a profit. They can make a profit by charging a commission. What changes need to be made to the quotas if they charge a commission of 10%?

Now it is your turn (or your students' turn)!
Do you have any ideas?
Please calculate the quotas so that the bookmaker gets to keep 10% of the total money wagered or placed.

We will take two different approaches here. The first approach involves deducting 10% of the total money placed, while the second approach will see us deducting 10% of the pay-out. In either case, the bookmaker gets to keep €6,000, which is a 10% commission.

If we deduct 10% of the money placed at the beginning, the amount the bookmaker now considers to be placed on each of the outcomes decreases by 10% to €9,000 (if Team A wins), €18,000 (if Team B wins) and €27,000 (if the game ends in a draw). What are the revised quotas in this case?

Outcome	Quota calculation	Quota
Team A win	$\frac{9{,}000}{9{,}000+18{,}000+27{,}000} = \frac{1}{6}$	6.00
Draw	$\frac{27{,}000}{9{,}000+18{,}000+27{,}000} = \frac{1}{2}$	2.00
Team B win	$\frac{18{,}000}{9{,}000+18{,}000+27{,}000} = \frac{1}{3}$	3.00

In this case the quotas remain unchanged. The implied probability $\left(\frac{1}{6} + \frac{1}{2} + \frac{1}{3}\right)$ still sums to 1. This should not come as a surprise. Your students can arrive at this conclusion themselves, and they might also consider if this holds true for other bets. Why is this? We introduce the keyword 'linear' at this point as a clue and move on to the second approach, that is to calculate 'backwards' starting from the pay-out.

In this case, instead of paying out €60,000, the bookmaker only intends to pay out €54,000, as he wants to gain €6,000. How will this be achieved? For all three outcomes, €54,000 is to be paid out. This will cause the quotas to change! What will the revised quotas be?

Outcome	Quota calculation	Quota
Team A win	$\frac{10,000}{54,000} = \frac{1}{5.4}$	5.40
Draw	$\frac{30,000}{54,000} = \frac{1}{1.8}$	1.80
Team B win	$\frac{20,000}{54,000} = \frac{1}{2.7}$	2.70

Now we have discussed two different approaches to find out what happens if the bookmaker wants to earn a commission. Which approach would you choose? Let us imagine how bettors would react, if they only received 2.7 times or 5.4 times the money they initially placed when the quota offered at the time of the bet was 3 or 6 in case of a win. The reply to angry customer enquiries would be that it said in the terms and conditions that 10% of profits would be retained. Such customers will probably opt for a different bookmaker the next time. As our initial examples show, bookmakers usually take the latter approach. They lower their quotas prior to posting them so that their commission is already deducted (invisibly). Only those performing some calculations themselves (or searching for relevant information on the web) can see which betting agency keeps the highest commission. From what we gathered, these commissions can be anything up to 10%—but due to stiff competition in the market it is often closer to 5%. In some cases the commission is even close to 0%. Here we see a great opportunity for some interesting homework: Ask your students to calculate the commission of various betting agencies for several matches.

5.5 Additional Question: Do Bookmakers Participate in Betting?

The time has come to review one of our modelling assumptions. We assumed that the bookmaker only wants to earn a commission based on the bets placed by their customers. However, would it not be advisable for the bookmaker to make use of his expert knowledge on the current form of the team to place bets himself? Using the figures from our last example we will consider what might happen if the bookmaker is convinced, from his study of the form and/or due to insider information, that Team A will win and so he intends to bet €20,000 on Team A to win.

A possible way of doing this (not the smartest way!) is for the bookmaker to become his own customer and place €20,000 on Team A to win. We assume that the bookmaker determines the quotas after all the bets have been placed and takes his commission based on the pay-out. We acknowledge that the idea that bettors would place money on an outcome without knowing the quotas is an unrealistic scenario in most countries but we will work with this simplified scenario for now to demonstrate some key concepts. Therefore, the effect of the bookmaker's wager on the quotas is easy to calculate. The total money placed rises from €60,000 to €80,000, his commission from €6,000 to €8,000. What will the resulting quotas be?

At this point you (and the students in your class) get another opportunity for a little exercise.
Please calculate the new quotas!

Here are our calculations with the commission deducted according to our second approach.

Outcome	Quota calculation	Quota
Team A win	$\frac{30{,}000}{72{,}000} = \frac{5}{12} = \frac{1}{2.4}$	2.40
Draw	$\frac{30{,}000}{72{,}000} = \frac{5}{12} = \frac{1}{2.4}$	2.40
Team B win	$\frac{20{,}000}{72{,}000} = \frac{5}{18} = \frac{1}{3.6}$	3.60

As a customer of this bookmaker, if you wish to calculate the commission based on these quotas we first find the ratio of the total money placed to pay-out, in our case:

$$\frac{30{,}000}{72{,}000} + \frac{30{,}000}{72{,}000} + \frac{20{,}000}{72{,}000} = 1.11$$

Then the commission is calculated by dividing 1 by our ratio and subtracting this answer from 1:

$$1 - (1 \div 1.11) = 0.1 \text{ or } 10\%.$$

If the bookmaker (or somebody else) places a €20,000 bet on Team A to win, this will make the previous outsider the new front-runner. The quota is reduced considerably. We now challenge you to try to find out how much money needs to be placed on a Team A win so that the quota falls to 1 or below due to the commission. This finding (where quotas fall in line with an increase in the money placed on an outcome) forces the bookmaker to place a limit on the amount of money that can be placed on a single bet.

Let us now return to our bookmaker and assume he will choose a different path when participating in betting himself. Do you have any ideas of an alternative option available to the bookmaker? We suspect that he is able to change the quotas if, or when, he chooses to place a bet.

> Which quota shall the bookmaker alter and in what way, so that he can secretly bet €20,000 on Team A to win?
> What will he gain, or risk, in doing so?

An answer may not present itself to us immediately. Consequently, we simulate the effects of a change in quotas on the totalisator (the bookmaker who determines the quotas after bets have been placed). We assume that he will alter one or more quotas after betting has closed but before the event starts. That way, he can secretly participate without openly placing any bets. To prevent discovery, he has to make sure that his commission of 10% can also be calculated from the changed quotas. How can we simulate such deliberations? A spreadsheet analysis can be used to carry out these calculations quite easily. Let us use the figures from our example once again, then change the calculated quotas and observe their effects on the three possible pay-outs for the match. Recall that €10,000 is placed on Team A to win, €20,000 is placed on Team B to win, and €30,000 is placed on a draw. The bookmaker also places €20,000 by changing the quotas. Let us start with our initial figures and the quotas calculated (Fig. 5.1).

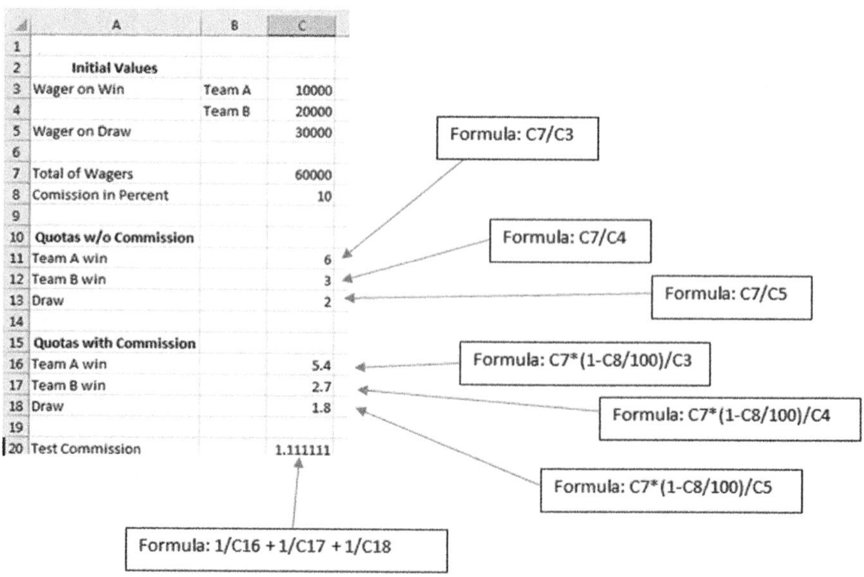

Fig. 5.1 Spreadsheet calculation of quotas with commission

As a result, 10% of the money placed remains as commission. For completeness we also list the pay-outs:

Pay-out	
54,000	Formula: C3 * C16
54,000	Formula: C4 * C17
54,000	Formula: C5 * C18

As expected, each of these pay-outs yields a commission of €6,000 for the bookmaker.

Let us now make our first attempt at changing the quota for a Team A win. Let us double the quota for a Team A win in our simulation. The quota will now be 10.8. How do the other quotas have to change, so that the test commission will remain at 1.1111? The implied probability for a Team A win decreases from 0.1852 (1 ÷ 5.4) to 0.0926 (1 ÷ 10.8) after doubling the quota—in essence the likelihood of a Team A win has been halved. If we want the total of all three implied probabilities to sum to 1.1111, we are short 0.0926. Therefore we need to add half of 0.0926 to each of the other two implied probabilities and decrease the quotas accordingly.

Altered implied probability for a Team A win	0.0926
Altered implied probability for a Team B win	(1 ÷ 2.7)+(0.0926 ÷ 2)=0.4166667
Altered implied probability for a draw	(1 ÷ 1.8)+(0.0926 ÷ 2)=0.6018519
Calculated total	1.1111

The recalculated quotas are:

Team A win	10.8
Team B win	2.4
Draw	1.6615385

The recalculated quotas are in line with our expectations. If one quota is to be increased randomly, the other quotas must be reduced 'accordingly' if we wish to maintain a fixed commission rate. However, the pay-out is what is really of interest to us. How will the pay-out change depending on the recalculated quotas?

Outcome	Calculation	Pay-out (€)	Bookmaker profit/loss (€)
Team A win	10.8 × 10,000	108,000.00	−48,000.00
Team B win	2.4 × 20,000	48,000.00	12,000.00
Draw	1.6615385 × 30,000	49,846.16	10,153.84

The random changes to our quotas have had a significant impact on the results. One thing we notice is that the bookmaker risks losing a lot of money, €48,000 to be exact, if Team A wins. This means that the bookmaker's first attempt at changing the quota for a Team A win was not advantageous.

We have several options for the next attempt at changing the quotas. For example, the bookmaker could increase the quotas for a Team B win or a draw and thus decrease the quota for a Team A win. The next step then would be to consider how we can facilitate the bookmaker's desire to place €20,000 at most on this event. We would also have to determine his maximum loss; is it the €20,000 he gambles, or is he only willing to lose his commission.

We will start by decreasing the quota for a Team A win (you are also invited to do your own experiments here). Which quota have you chosen to alter? We decided to half the original quota, hence our new quota for a Team A win is 2.7. Our spreadsheet analysis then provides us with the following results:

Altered implied probability for a Team A win	$1 \div 2.7 = 0.3704$
Altered implied probability for a Team B win	$1 \div 2.7 - [(0.3704 - 0.1852) \div 2] = 0.277778$
Altered implied probability for a draw	$1 \div 1.8 - [(0.3704 - 0.1852) \div 2] = 0.462956$
Calculated total	1.1111

The recalculated quotas are:

Team A win	2.7
Team B win	3.6
Draw	2.16

The pay-out associated with each of the revised quotas is:

Outcome	Calculation	Pay-out (€)	Bookmaker profit/loss (€)
Team A win	2.7 × 10,000	27,000.00	33,000.00
Team B win	3.6 × 20,000	72,000.00	−12,000.00
Draw	2.16 × 30,000	64,800.00	−4,800.00

This time we are on the right track. If the quota for a Team A win is lowered, the bookmaker's profit will increase if Team A actually wins. In addition to his commission of €6,000, he earns €27,000. This however comes with some financial risks. Should Team B win, then he will lose €12,000 and in the case of a draw he will lose €4,800.

However, one question still remains unanswered: it still needs to be decided whether to limit the bookmakers maximum loss to €20,000 (the sum he is willing to gamble), or whether the commission should be the cap on his loss. To follow this train of thought, we need to find a way to incorporate this idea into our spreadsheet analysis so that we can invert our thinking. The question now is how can quotas be calculated when given a maximum loss that the bookmaker is willing to incur?

5.6 Comment on Teaching Methods

At this point a decision relating to our mathematics class has to be taken. One possible approach that would allow us to progress is 'systematic' trial and error. Using the spreadsheet analysis, we could try to find a suitable value. The other approach would be to merge the functions we used for making our 'forward' calculation into one single function (which can hopefully be inverted) and then analyse it. In some cases there is a third option—using GeoGebra, or a similar software package, you can create a graph of your function and then consider the output. Since our aim is not to find a result that is precise from an analytical point of view, but is instead to find a quota accurate to within two decimal places, the second method will suffice.

We will now attempt to find an answer to our problem with the help of functions. Essentially we must find a new, random quota for a Team A win that is not conspicuous (i.e. the total of the implied probabilities for all three betting options still has to sum to 1.111111 whilst remaining somewhat realistic). In addition to this, the loss incurred by the bookmaker when the quotas are changed has to be limited to €20,000, his secret stake.

What is the first step in looking for a suitable function? Is it immediately obvious what we should do? First, we will test the obvious suggestions: similar to before, we lower the quota for a Team A win by a certain value, and at the same time increase the quota for a Team B win by the same value. Here are the quotas before the manipulation as a reminder:

Outcome	Quota	Implied probability
Team A win	5.40	0.185185185
Draw	1.80	0.555555556
Team B win	2.70	0.37037037
Calculated total		1.111111

The total sum of the implied probabilities is 1.111111 and this is the total that must be maintained after changing the individual quotas.

Let us now use the approach outlined above where we lower the quota for a Team A win by a certain amount and in line with this increase the quota for a Team B win by the same value. How about using a value such as 2 or 3? We will begin by testing a decrease in the quota for a Team A win by 2 and a subsequent increase in the quota for a Team B win by 2. The new quotas and implied probabilities are:

Outcome	Quota	Implied probability
Team A win	3.40	0.294117647
Draw	1.80	0.555555556
Team B win	4.70	0.212765957
Calculated total		1.06243916

The total sum of the implied probabilities is now 1.06243916—which is not the same as before. If we change the quota by 3 instead of 2, it does not work either. From these attempts we can conclude that this approach will not work and so we reject it.

Instead of increasing and decreasing our quotas, we will instead look at reducing and increasing our quotes by a set proportion. For example, what if we divided the current quota offered for a Team A win by a certain value, such as 2, and in exchange increase the quota for a Team B win by same proportion.

Outcome	Quota	Implied probability
Team A win	2.70	0.37037037
Draw	1.80	0.555555556
Team B win	5.40	0.185185185
Calculated total		1.111111

The total sum of the implied probabilities is again 1.111111—perfect. Are we done now? When teaching proof we would have encountered the idea that one (counter) example, will suffice to show that an idea is incorrect, but plenty of examples do not prove that an idea is correct. For example, an infinite number of uneven numbers are prime numbers, but not all of them are. So what do we do now? Let us perform another calculation, this time with 3 as our scaling factor. Here are the new quotas and expected profits:

Outcome	Quota	Implied probability
Team A win	1.80	0.555555556
Draw	1.80	0.555555556
Team B win	8.10	0.12345679
Calculated total		1.234567901

The total sum of the implied probabilities this time around is 1.234567901—which is incorrect. Once again our idea has been shown to be flawed and we (in the role of our eager students) need to reconsider how we can reach our goal. How does a decreased quota for a Team A win correlate with an increased quota for a Team B win, if the total sum of the implied probabilities has to remain the same? To gain more insight into this problem we write down the following word equations:

Initial total sum of the implied probabilities = 1.111111 = implied probability for a Team A win + implied probability for a Team B win + implied probability for a Draw

New total sum of the implied probabilities = 1.111111 = (decreased) implied probability for a Team A win + (increased) implied probability for a Team B win + implied probability for a Draw

If we combine both equations, the following remains:

Implied probability for a Team A win + Implied probability for a Team B win = (decreased) Implied probability for a Team A win + (increased) Implied probability for a Team B win.

In order to solve this equation the first question we must ask at this point is "How do we express 'decreased' and 'increased' in mathematical form in the equation?" In our first attempt at solving this problem we sought to decrease the quota for a Team A win by x and increase the quota for a Team B win by the same amount. The aforementioned equations can now be written as:

$$\frac{1}{quota\ A} + \frac{1}{quota\ B} = \frac{1}{quota\ A - x} + \frac{1}{quota\ B + x}$$

When we look at this equation, and based on our previous numerical calculations, we are not entirely convinced that it is correct. How can we be certain that the amount added to the quota for a Team B win should be the same amount taken from the quota for a Team A win? If we subtract a different value, let's say y, from quota B instead, we get an equation with two unknowns. How can we go about identifying these unknowns? It is likely that when given such an equation students would initially insert the known quota values and try to solve the equation for x, instead of considering the questions we have just raised. Let us see if this approach works, and even if it does not, maybe we can learn something from it.

$$\frac{1}{5.4} + \frac{1}{2.7} = \frac{1}{5.4 - x} + \frac{1}{2.7 + x}$$

x will certainly not be 5.4, otherwise the new quota for a Team A win would equal zero—something that does not occur in real life. Using algebra, we will get a much simpler equation:

$$x^2 - 2.7x = 0$$

with the solutions $x = 0$ and $x = 2.7$. This appears to make sense (in other words, we could have realised it right away). If we alter both quotas by 2.7, then we end up exchanging them. This means the original quota for a Team A win now becomes the quota for a Team B win and vice versa.

What effect does this have on pay-outs?

Outcome	Calculation	Pay-out (€)	Bookmaker profit/loss (€)
Team A win	2.7 × 10,000	27,000.00	33,000.00
Team B win	5.4 × 20,000	108,000.00	−48,000.00
Draw	1.8 × 30,000	54,000.00	6,000.00

As a preliminary conclusion, we can see that as eager students, using step-by-step considerations and calculations, we have found a way for the bookmaker to change the quotas. This enables him to secretly participate without anyone noticing (as profits can be recalculated using the quotas). However, he might lose up to €48,000, much more than he is prepared to risk. So what should we do?

Two options present themselves to us: if we share the change in quota between a Team B win and a Draw, we can significantly reduce the potential maximum loss. However, on the other hand if we choose to stick with our decision to only change the quotas for a Team A win and a Team B win, we need to start with our result and then work backwards. To do this we need to ask ourselves: *For which quota for a Team B win is the loss €20,000 at the most? What is the corresponding quota for a Team A win?* Once again we have to make a decision: which of the two options should we implement? As with all decisions in the course of this project, we strongly advise letting the class discuss the questions posed and allow them to make the important decisions. This is the only way students can acquire the skills needed for fulfilling the educational goal of 'learning for real-life'. We prefer the option that requires us to find suitable quotas for a Team A win and a Team B win when assuming a maximum loss of €20,000. Once again we will approach the problem using algebra, and thus try to create suitable equations.

What does the quota for a Team B win have to be, so that the pay-out turns into a loss of just €20,000? To do this we would need to solve the following word equation using our particular values:

Money wagered on a Team B win multiplied by the quota = pay-out.

In this particular example, this equation can be rewritten as:

$$20{,}000 \times q = 80{,}000$$

Hence we know the quota, $q = 4$. The quota for a Team B win can be at most 4. Now we must calculate the corresponding quota for a Team A win. Previously we used the following equation:

1.111111 = (decreased) implied probability for a Team A win + (increased) implied probability for a Team B win + implied probability for a Draw

Let us now insert the quota value we just determined for a Team B win:

$$1.111111 = y + \frac{1}{4} + \frac{1}{1.8}$$

Therefore $y = 0.305555444$. Thus, the new quota for a Team A win is

$$\frac{1}{0.305555444} = 3.27.$$

In conclusion, with a bit of perseverance students in lower secondary school have now reached their goal. They are now aware of how bookmakers can place bets by adjusting the quotas in the opposite direction than the legitimate bets placed would suggest. We must now ask the following question: *what happens if the altered quotas for a Team B win suddenly trigger additional bets*? This means a sudden surge in risk for the bookmaker. As it is more likely that Team B wins, he has to shoulder this loss himself. Bearing this in mind, it makes perfect sense for the bookmaker to fix his quotas in a way so that he always wins, no matter what the outcome. The commission helps him to do so. If he succeeds, people say 'the books are balanced'. The bookmaker can calmly await the outcome of the match because he will make some money regardless.

It is up to the students (with a little encouragement from their teacher) to decide whether they want to make an attempt at generalisation, by looking for appropriate formulae or functions. We deliberately do not press for an answer here and instead we take matters one step further than the totalisator by looking at the case of a conventional bookmaker. We claim it is in his best interest that his quotas resemble the actual bets placed as closely as possible, thus enabling him to cut his losses. Below is a modelling exercise we set up to put this theory to the test.

5.7 The Bookmaker's Fear of Risk

We already pointed out that most bookmakers offer their quotas while betting is ongoing. For example, if somebody places a bet at a certain time, he/she has to receive his/her pay-out based on the quota that was on offer at that particular time. This might pose a risk to the bookmaker if the betting behaviour changes and the initial quotas no longer match the current incoming bets. In such cases, the bookmaker himself has two options: he places a bet himself or he agrees to accept the risk. We assume that he does not want to participate in betting and so we ask ourselves, how can he minimise the risk?

If your students are still in doubt over this matter we will try to convince them, and you, with the following example. On August 2nd, 2017, Paddy Power offered the following quotas for the first English Premier League match of the season between Arsenal and Leicester City which was to take place on the 11th of August; the quota for an Arsenal win was 1.36; the quota for a draw was 5.00 and the quota for a Leicester win was 8.00.

The season before, Arsenal finished 6th in the league with a total of 75 points and only 9 losses over the course of the season, whereas Leicester finished 12th amassing a total of 44 points and recording 18 losses over the season. As such, Arsenal was the favoured team in this match. Let us assume that as of August 2nd, €10,000,000 had been wagered on this match with Paddy Power. Moreover, let us also assume that the quotas mentioned are roughly proportional to the money placed on each outcome.

> Determine the nature of the bets placed so far—how much was placed on an Arsenal win, a Leicester win or a draw?

Once again this is a new concept. You (and your students) are supposed to find out how much was placed on each of the three possible outcomes. To do this, you will need to reverse the procedure used previously. We would advise letting your students figure this out for themselves, however we will offer a possible approach. As always, we do a little mental arithmetic to see what the result might be. The largest share of bets is in favour of an Arsenal win, with a small share placed on a draw, and a smaller share again placed on a Leicester win. At first, we determine how much commission the bookmaker takes.

The total sum of the implied probabilities is:

$$\frac{1}{1.36} + \frac{1}{5} + \frac{1}{8} = 1.060294118,$$

just a little above 1. This means that the bookmaker has set his commission at around 6% (i.e. €600,000). It is worth noting that bookmakers generally choose not to set their quotas below 1 on purpose. This ensures that a bettor is guaranteed to have some return on their wager in the case of a win.

Continuing with this problem, how is the remaining €9,400,000 (10,000,000–600,000) distributed among the three possible outcomes? Let us remember the basic principle behind sports betting. The winners receive all the money (minus the commission). Therefore, we can calculate the wager placed by inverting the quota. €9,400,000 $\times \frac{1}{1.36} =$ €6,911,764.71 and so this was the amount of money placed on an Arsenal win. Likewise, €9,400,000 $\times \frac{1}{5} =$ €1,880,000 was the amount of money placed on a draw and so the remaining €1,175,000 (9,400,000 $\times \frac{1}{8}$) is the amount placed on a Leicester City win (provided the bookmaker calculates as a totalisator would). The recalculated total of wagers placed is therefore €9,966,764.71, which is a little under the 10 million euro. We can conclude that for ease of presentation the quotas were rounded a little.

5.8 Accounting for the Unexpected: A Slight Variation to Our Model

Let us now consider the following virtual situation: We assume something out of the ordinary, for didactic reasons, to illustrate the risks a bookmaker can take if he participates in betting, for whatever reasons. In this case, we are going to alter the implied probabilities. We will assume that rumours emerged that something in the Arsenal team lunch has caused food poisoning. Most players and substitutes were rushed to the hospital and were prevented from playing (this is only an assumption for the sake of the modelling activity: in reality Arsenal won 4–3 on the night). As rumours would have it, a medley of substitutes and trainees would play instead. The instant these rumours, which were later confirmed by the hospital and news agencies, spread across the internet the distribution of the money wagered changed considerably. Suddenly, Leicester City stood a good chance of winning. An extra €20,000,000 was suddenly placed on Leicester City to win (in our assumed scenario). Assume that the bookmaker does not alter his quotas and that Leicester City actually won the game. How much of a pay-out does the bookmaker have to make? This is a little exercise in working with large numbers: €21,175,000 × 8 = €169,400,000. If a little over €21 million was placed on Leicester City to win, the bookmaker pays approximately €170 million to the winners. In this case, the €8,791,764.71 previously wagered on an Arsenal win or a draw, that he gets to keep, offers him little consolation.

We can now agree that bookmakers want to avoid such risks as the one outlined here while at the same time avoiding having to place large bets themselves. How can bookmakers achieve this? Do you have any ideas? Perhaps they might try to manipulate the outcome of the match? Yes, this seems to happen occasionally but it is illegal and is known as betting fraud. So what legal options does the bookmaker have? Let us perform a little modelling calculation. The extensive losses incurred in our last example are due to the fact that the bookmaker did not adjust his quotas to reflect the altered situation and the new wagers placed. Therefore, the underlying principle for the bookmaker is to constantly adjust his quotas to reflect ongoing wagers and those already placed. This means constantly recalculating the quotas for the current level of wagers placed, as a totalisator would. The initial quotas will be cautious estimates. Now the question to be answered is, how should he constantly update his quotas? What would your suggestions be?

5.9 Modelling Calculations Based on the Bookmaker's Reaction

In this section we are going to model several courses of action. We shall use a simplified version of reality for our model: we assume that each bet placed amounts to a certain sum, for example €10. In reality, the amounts wagered typically vary from €1 up to the applicable limit. Without this simplification we would have to employ complicated statistics to eventually arrive at the same result.

Rather than working with extremely large numbers that may complicate matters, we will use a simpler example whereby we assume €10,000 euro has been wagered in total. After deducting our 10% commission, we know that €3,000 has been wagered on a Team A win; €5,000 on a Team B win and €1,000 on a Draw. The three quotas can now be calculated as follows:

Outcome	Quota calculation	Quota
Team A win	$\frac{3000}{9000} = \frac{1}{3}$	3.00
Draw	$\frac{1000}{9000} = \frac{1}{9}$	9.00
Team B win	$\frac{5000}{9000} = \frac{1}{1.8}$	1.80

This would be an ideal situation. No matter what the outcome, the bookmaker earns his commission, in our example €1,000.

Now suppose an additional €10,000 in bets is added to our model. The first thing to consider is which outcome they are wagered on. To assist us in exploring this scenario we will carry out a simulation. In our first simulation, the new bets (a total of €10,000) are placed in the same ratios as the previous ones. This means that after deducting the 10% commission, an additional €3,000 is placed on a Team A win, €5,000 on a Team B win, and €1,000 on a Draw. What does the bookmaker have to do in this simulation? Nothing! In this case the quotas remain the same while the pay-outs and the money received by the bookmaker doubles.

We will now explore some extreme cases. For example, we will first assume that all additional wagers are placed on the one outcome, a Team A win. What happens in this case? Please calculate the new pay-out assuming the quotas will remain the same. What have you come up with? Our result is as follows:

Outcome	Calculation	Pay-out (€)	Bookmaker profit/loss (€)
Team A win	$3.0 \times 12{,}000$	36,000	−16,000
Draw	$9.0 \times 1{,}000$	9,000	11,000
Team B win	$1.8 \times 5{,}000$	9,000	11,000

The pay-outs indicate that in this case the bookmaker has taken a gamble. Depending on the outcome, he will either win or lose. Let us now calculate what happens if he recalculates the quotas every time 100 new wagers (with each individual wager amounting to €10) have been placed. We also have to bear in mind that pay-outs need to be calculated based on the quota valid at the time the bet was accepted. We need to calculate different pay-outs according to the time the wager was actually placed, and then add them all up. Let us create a chart for this purpose, using these initial values. The original sum of money placed was €10,000 with a commission of €1,000.

Outcome	Calculation	Pay-out (€)	Bookmaker profit/loss (€)
Team A win	3.0 × 3,000	9,000	1,000
Draw	9.0 × 1,000	9,000	1,000
Team B win	1.8 × 5,000	9,000	1,000

Now suppose another 100 bets amounting to €1,000 is placed on a Team A win. The total of all bets placed is now €11,000 while the commission is €1,100. We will now update our tables.

Outcome	Money wagered (€)	Quota calculation	Quota
Team A win	3,900	$\frac{3900}{9900} = \frac{1}{2.53846}$	2.53846
Draw	5,000	$\frac{1000}{9900} = \frac{1}{9.9}$	9.90
Team B win	1,000	$\frac{5000}{9900} = \frac{1}{1.98}$	1.98

Outcome	Calculation	Pay-out (€)	Bookmaker profit/loss (€)
Team A win	2.53846 × 3,900	9,900	1,100
Draw	9.90 × 1,000	9,900	1,100
Team B win	1.98 × 5,000	9,900	1,100

In this instance the bookmaker managed to earn a commission of exactly 10%. In reality this may not always be the case. This is because bets are laid at different times, at different quotas and by solely basing your commission calculations on the most recent quota, your calculations do not take into account earlier different quotas. As a result you generally cannot guarantee that the bookmaker will get exactly 10% commission.

In the previous example, you will notice that we calculated the quotas and pay-out as if they had been recalculated after betting closed. However, this is not generally the case. We have decided to calculate what happens if the bookmaker recalculates the quotas every time 100 new wagers have been placed. Hence, the quota will not change for the first 100 additional bets. The new quotas calculated in our table will only apply from the 101st bet onwards. Here are the corrections to the table:

Outcome	Calculation	Pay-out (€)	Bookmaker profit/loss (€)
Team A win	3.00 × 3,900	11,700	−700
Draw	9.00 × 1,000	9,000	2,000
Team B win	1.80 × 5,000	9,000	2,000

When we consider these results, one trend becomes apparent which we have also come across as a deterrent in our extreme case above. The bookmaker is still taking a risk in this scenario. A table follows with all the calculations for an additional €10,000 of wagers, that is an additional 1,000 bets. Remember that to determine the quota (and the pay-out!) for each subsequent 100 wagers we use the amounts bet on the previous 100 wagers. For example, in the case of the second additional €1,000 that is wagered (recall that the bookmaker will take his 10% commission at the start) only €900 is actually bet on a Team A win. In this case the quota and pay-out are determined as follows:

Outcome	Money wagered (€)	Quota calculation	Quota
Team A win	3,900	$\frac{3900}{9900} = \frac{1}{2.53846}$	2.53846
Draw	5,000	$\frac{1000}{9900} = \frac{1}{9.9}$	9.90
Team B win	1,000	$\frac{5000}{9900} = \frac{1}{1.98}$	1.98

Outcome	Calculation	Pay-out (€)	Bookmaker profit/loss (€)
Team A win	11,700+(2.53846 × 900)	13,984.61	−1,984.61
Draw	9.00 × 1,000 (9.90 × 0)	9,000	2,000
Team B win	1.80 × 5,000+(1.98 × 0)	9,000	2,000

This procedure is repeated for each additional €1,000 that is wagered and the results are tabulated here for you.

Table 5.1 Changes in quotas for every additional €1,000 wagered

	Team A win	Team B win	Draw
First 10,000	3000	5000	1000
Payout	9000	9000	9000
Result	1000	1000	1000
Additional 1,000	3900	5000	1000
Payout	11,700	9000	9000
Result	−700	2000	2000
Additional 2,000	4800	5000	1000
Payout	13984.61	9000	9000
Result	−1984.61	3000	3000
Additional 3,000	5700	5000	1000
Payout	16009.61	9000	9000
Result	−3009.61	4000	4000
Additional 4,000	6600	5000	1000
Payout	17856.98	9000	9000
Result	−3856.98	5000	5000
Additional 5,000	7500	5000	1000
Payout	19575.16	9000	9000
Result	−4575.16	6000	6000
Additional 6,000	8400	5000	1000
Payout	21195.16	9000	9000
Result	−5195.16	7000	7000
Additional 7,000	9300	5000	1000
Payout	22738.02	9000	9000
Result	−5738.02	8000	8000
Additional 8,000	10,200	5000	1000
Payout	24218.67	9000	9000
Result	−6218.67	9000	9000
Additional 9,000	11,100	5000	1000
Payout	25648.08	9000	9000
Result	−6648.08	10,000	10,000
Additional 10,000	12,000	5000	1000
Payout	27034.57	9000	9000
Result	−7034.57	11,000	11,000

Inserting all these numbers into Table 5.1 is an arduous task, especially if you ask your students to perform these calculations when quotas are to be recalculated every 10 bets, or even whenever a new bet is placed. It requires quite a bit of effort, so we wonder if it would not be better to involve a little more mathematics, maybe a formula. If we could derive a formula, a spreadsheet analysis would do the rest of the work for us.

When analysing the numbers listed in Table 5.1, we notice that the bookmaker still takes a risk using this approach, despite the adjustments made to the quotas. He may lose up to €7,000, and remember, for matches with a higher turnover it may be even more. However, at least the risk is lower than before, when he maintained the initial quotas throughout. You will recall the loss in the extreme case amounted to €16,000. Adjusting the quotas every 100 additional wagers, or for every €1,000 placed, resulted in the risk taken by the bookmaker being reduced by more than half. We might also conclude that more frequent updates of the quotas might help to cut the losses even more.

If we wish to test this assumption, we either have to be very diligent or use other mathematical means. As students are often told during their studies at university, mathematics is said to be an invention of people who prefer thinking as opposed to performing calculations. So rather than writing down plenty of mathematical calculations, we would like to consider other approaches your students can pursue. What advice would you give to your students, if they asked for it?

We first suggest writing down in detail what we want to achieve and what we already know. A spreadsheet analysis will help us in this situation. We can create the spreadsheet so that we only have to input the new total amount wagered on each bet and the program will then calculate the commission, the new quotas, the expected pay-out and the financial result for the three possible outcomes.

We already know how to calculate the commission and we can calculate the quotas from the bets placed and the expected pay-out from the quotas. We can then compare the expected pay-outs to the bets placed to obtain our financial result, that is, the bookmaker's profit/loss. In Table 5.2 we again begin with €10,000 initially placed and we revise our quotas each time a new bet is placed.

We have started inserting our initial values and the formulae for the changes into a table. The values in our table appear to be correct, but if we continue on like this

Table 5.2 Changes in quotas for every additional €10 wager

	Team A win	Team B Win	Draw
First 10,000	3000	5000	1000
Quota	3	1.8	9
Pay-out	9000	9000	9000
Profit/Loss	1000	1000	1000
Additional 10	3009	5000	1000
Quota	3	1.8	9
Pay-out	9027	9000	9000
Profit/Loss	983	1010	1010
Additional 10	3018	5000	1000
Quota	2.99	1.80	9.01
Pay-out	9053.91	9000	9000
Profit/Loss	966.09	1020	1020

Table 5.3 Pay-out on Team A win and profit/loss calculations

Number of additional deposits	Pay-out on Team A win (€)	Profit/Loss (€)
1	9,027	983
2	9,053.95	966.05
3	9,080.84	949.16
4	9,107.68	932.32
5	9,134.46	915.54
6	9,161.20	898.80
7	9,187.88	882.12
8	9,214.51	865.49
9	9,241.09	848.91
...	...	...
95	11,361.55	−411.55
96	11,384.56	−424.56
97	11,407.53	−437.53
98	11,430.48	−450.48
99	11,453.39	−463.39
100	11,476.26	−476.26
...	...	...
996	26,270.48	−6,310.49
997	26,284.00	−6,314.00
998	26,297.51	−6,317.51
999	26,311.02	−6,321.02
1000	26,324.52	−6,324.52

until the 100th additional bet, the table gets much too large and the task becomes very tedious. We must ask ourselves if there is another way. Do you have any ideas?

The only data we are concerned with at this point relates to the pay-out for a Team A win and the ensuing financial result. In Table 5.2 we calculated the pay-out by multiplying the new quota (determined from the previous total amounts wagered) with the new amount wagered and then added this to the previous pay-out. The profit/loss is then calculated by subtracting the pay-out for a Team A win from the total amount wagered (including commission). We can conclude from the first two bets that the bookmaker will not do well if, after a certain time, additional bets are all placed on a single outcome. Even if he reacts swiftly (and adjusts the quotas accordingly), his profit will go down if this scenario becomes a reality.

We shall employ some more mathematical methods and put in some more effort to make this table easier to understand. To this end we focus only on three factors: number of additional deposits (bets), the pay-out and the resultant profit/loss incurred by the bookmaker. Table 5.3 was imported from Excel.

Let us now compare the figures for 100 or 1000 additional deposits (i.e. bets placed on a Team A win) in an attempt to identify the optimum time to recalculate

Table 5.4 Pay-out on Team A win and profit/loss calculations after 100 and 1,000 bets

approach used	Pay-out for Team A win after 100 bets (€)	Profit/Loss (€)	Pay-out for Team A win after 1000 bets (€)	Profit/Loss (€)
Fixed quota	11,700	−700	36,000	−16,000
Change after 100 bets	11,700	−700	27,034.57	−7,034.56
Change after each bet	11,476.26	−476.26	26,324.52	−6,324.52

the quotas. We notice that these frequent alterations help cut the losses incurred by the bookmaker, but they cannot prevent them altogether.

Please find the new values summarised in Table 5.4.

For quite some time now we have dealt with the bookmaker's fear of risk. Our motivation here was not to express our sympathies for this line of work, but instead we were looking for an interesting example to help us illustrate the modelling cycle. Your active participation made it easier for you to grasp how a problem develops, how new issues arise that need to be dealt with, and they in turn lead to new questions. In our example we get a clear answer regarding the need for all this work and we can conclude that the effort was definitely worth it. Bookmakers who use a computer program to calculate and adjust their quotas based on the latest bets, can minimize their losses in extreme cases, for example if all wagers are placed on one outcome, and keep their books almost balanced while all the time obtaining their desired commission, no matter what the outcome of the match may be.

5.10 Additional Modelling Assumptions and Simulations

In this chapter of the book we have looked into the topic of sports betting in detail. Therefore, you (or your students during the brainstorming session at the close of this unit) might think of other risky situations for the bookmaker, even if he continuously keeps updating his quotas. The following two types of risk come to mind:

The favourite wins

When Team A is the favourite, all the bettors place their money on a Team A win and nobody on Team B win or a Draw. If the bookmaker started with a certain quota, let us say a cautious 1.5 for a Team A win, and he accepted 10 bets on a Team A win, his profit is €100 and the pay-out amounts to €150 (a loss of €50). The totalisator on the other hand deducts his 10% commission (his profit is €10), fixing the quota at 0.9. Those who bet on the right outcome, receive a final pay-out of €9 for their €10 invested (All those who bet will be thrilled).

The initial quota is far off the mark and hardly any bets are placed

When the bookmaker determines his opening quota, he obviously cannot calculate it on the basis of bets received, as no bets have been placed yet. In reality the bookmaker lets his experience guide him: what were the quotas last time for this type of match? Which team is the favourite and to what extent, according to the experts? Teams typically perform better at home—and so on. Once a wager has been accepted, a pay-out has to be made at the quota applicable at that time in case of a win. Let us assume that a bookmaker uses the following quotas (Team A win: 2; Draw: 3.33; Team B win: 5). Two people place additional bets for €10,000 each on a Team B win, one person places €10,000 on a draw and one bets the same amount on a Team A win. There are no other bets. Then apparently the quotas are incorrect. Should Team A win, the additional wagers of €20,000 will be paid out and another €20,000 can be gained—perfect for the bookmaker. In the case where the match ends in a draw, the pay-out will be €33,333 and the bookmaker gets to keep €6,667. Should Team B win, the prospects are dire: €50,000 has to be paid out twice. Consequently, the bookmaker will be at a loss of €60,000.

5.11 Didactic Considerations on Using the Topic of Sports Betting in Mathematics Class

How can course units on the topic be created in a way that students find them interesting and motivating? Some years ago we came up with the proposal presented below, which has been put to the test several times. But first please read the following two introductory remarks. Even though the topic of sports betting tends to get a lot of media coverage, this does not automatically make it an interesting topic for mathematics education. Students' own research often does the trick here instead of motivating explanations on the teacher's part. To this end we suggest simulating a sports match including a betting shop, where students can place their own bets using pretend money. This should also make it easier for students to understand why quotas cannot easily be determined by rolling a die. Quotas mirror bettor's expectations of the outcome. They thus differ from the expected results determined by a random generator. Despite our ardent hopes for a certain number or card during a game of dice or card game, the random generator remains unaffected by this (provided it is a real one and no cheating takes place). Our hopes however do affect the quotas—even intentionally (as we have shown during modelling calculations on the behaviour of the bookmaker). Because fans tend to support their own team and place wagers on them to win, even though this outcome is unlikely, this can result in the odds for an outsider winning dropping and the odds for the frontrunner steadily increasing. Students should realise in this case that even though we speak of expected profits when we explain quotas, the quotas are a result of the personal expectations of the people who place bets and are thus subjective.

Fig. 5.2 Game simulation

Students of all ages are always delighted when such experiments and meaningful 'games' bring colour to the 'dull' routine of school life. Unfortunately, this sort of 'colouring' is still all too rare in mathematics education. Our proposed course unit shall provide a welcome change here: a simulated sports match including sports betting. At first students should consider how they can incorporate a random match suited to betting into their mathematics education. We suggest soccer here although at school it may prove difficult if several groups of students form teams and then want to conduct live matches to determine the winner and to see if betting makes sense. The main impediment here is the duration of each match and the difficulty in judging the teams without any trials. After all, teams formed in a class lack an entire season's practice and a series of match results making it easier to determine a frontrunner.

Therefore, we suggest a simulation of penalty kicks on a table. In order to do this, students only need a certain degree of dexterity and the cap of a plastic bottle to act as the ball. 'Goal posts' are set up on a table using pens or books (see Fig. 5.2). Two players then take turns as 'penalty-takers' and try to get the cap into the goal by flicking it with their fingers. As there are no goalkeepers, success only depends on the direction and speed. Each team (in our simulation each penalty-kicker) gets 5 shots. If the result after this is still a draw, both kickers get another try until the match is decided (that is, one scores a goal and the other one does not). In doing so there will always be a winner, a draw is not possible. This makes the calculations easier. At a later stage rules may be changed to achieve a more realistic simulation. A little experimenting will tell if a goal 10 cm wide or 15 cm wide works better for a particular class, or if the distance between the penalty-taker and the goal should be one or two desks long. In case of two desks, it also depends if both desks are exactly on the same level—a difference in height is a handicap for the penalty-takers.

5.12 Trial Run

The teacher can initiate a test run for the planned tournament and the betting process. Two students volunteer as penalty-kickers and a betting shop is set up. The two volunteers hold a public training session that ends after ten kicks. Records are kept (for example student A scored 5 goals and student B scored 7). Now the betting shop begins its work: All students place their bets—except for the two penalty kickers to avoid betting manipulation. All students place 10 units of token coins to make calculations easier. If we assume that the class consists of 32 students, then 30 bets amounting to 300 units of token coins will be placed. Students may place their bets following their instincts or guided by reason, randomly, or by deliberate calculation. The teacher should refrain from exerting influence, just have them list who bet on whom.

You can probably guess how things will proceed from here: after wagers have been placed and the match ended, the designated bookmakers pay out profits to those who guessed right. Who receives how much token money? Based on this simple example students can possibly find out themselves what quotas are and how they can be calculated. As soon as the foundations have been laid, more complex situations (as the ones described previously) can possibly be dealt with and understood.

We should recognise that there are social issues associated with betting and gambling generally, such as betting fraud and gambling addictions. None of these behaviours are desirable but legal gambling and betting fraud can be analysed using mathematical means. On the other hand, addictions cannot be prevented or reduced by mere information or reprimands, a person's resolve is key. However, it is crucial in our view that students immerse themselves in this topic and consider the possible influences on their behaviour.

Chapter 6
How Do Experts Model? Using This Knowledge and Understanding in the Mathematics Classroom

In this chapter we turn our attention to expert modellers. At the start of our journey together in this book we appealed to your, and your students', natural capacity for modelling. Then we built on this capability using examples to make aspects of the modelling process explicit and learnable. We hope your students are now good novice modellers and that you are an even more capable modeller. By looking at what expert modellers do you can improve your modelling skills and make your mathematics classes more interesting and challenging.

Are you aware of the fact that thousands of people earn their living by developing mathematical models? It is reasonable to think of these people as expert modellers. However, you may ask: What exactly do these people do? What types of models do they develop and what kind of results does their work produce? We consider these questions in this chapter and, as in previous chapters, we ask for your active participation, even though we provide the necessary background information and guidance here. How will this benefit you and your students? We suggest you take ideas and materials for student projects or presentations from this chapter. As we would like you to better understand how easy or difficult it is to gather information for such presentations, we now ask you to do so yourself.

What can you find on the web about mathematics-in-industry or technomathematics?

Before embarking on our quest, we have a few more suggestions for you. We just introduced two keywords in the question we posed: mathematics-in-industry and technomathematics. Providing keywords is a useful teaching method for assigning topics for presentations, and using carefully selected keywords may help to focus the search. Such suggestions will help you direct the work of your students.

It is important to use carefully selected keywords. For example, if you enter the term industrial mathematics into your search engine, you will more than likely be directed to a specialisation offered at one of a number of universities

J. Maaß et al., *Mathematical Modelling for Teachers*, Springer Texts in Education, https://doi.org/10.1007/978-3-030-00431-6_6

worldwide, such as Johannes Kepler University in Linz (Austria) or other universities in your own country. Alternatively, you may be directed to colleagues active in this field, or to companies that benefited from collaboration with the aforementioned researchers/institutes. You may also discover others offering courses or doing research in industrial mathematics, for example Pennsylvania State University (U.S.), the University of Hamburg (Germany), University of Oxford (U.K.) or University of Limerick (Ireland).

Technomathematics on the other hand is linked to the activities of Professor Neunzert of the University of Kaiserslautern. The website describes the specialisation in technomathematics as follows: *The technomathematics specialisation focuses on modelling, analysis and numerical simulation of differential equations, and also on their optimisation and control* (http://www.mathematik.uni-kl.de/en/research/industrial-mathematics/). This important branch of applied mathematics deals with problems taken from physics, engineering, science, or the life sciences. It then develops new models and techniques for processing them analytically and numerically. Technomathematics-related research focuses on applied topics and is often influenced by problems the various departments of the Fraunhofer-Institute for Technomathematics and Mathematical Economics (ITWM) are experiencing. The mathematical methods and algorithms they develop are successfully applied to address problems faced by production companies.

The mathematical expertise of the technomathematics specialisation consists of modelling, analysis and simulation of (stochastic) partial differential equations and differential algebraic equations which will then be applied to areas such as fluid dynamics, as well as the life sciences, radiation, transport, models of traffic situations, semigroup equations to name but a few. Stochastic differential equations, kinetic models and non-local integro-differential equations are analysed in addition to classical elliptic, parabolic and hyperbolic equations. Moreover, methods of optimisation with partial differential equations are applied to solve inverse problems and optimal control problems for the topics mentioned above. This research area also forms a link between algebraic theory and control theory and its applications to electrical engineering and vehicle simulation.

We just presented you with a rather long description of technomathematics so that we can brainstorm together about what your students might make of this description. For instance, we noticed that the term 'modelling' was used from the outset. We take this as proof that experts actually perform mathematical modelling and naturally we are anxious to see what exactly they model. You will probably be familiar with some of the terms mentioned such as, analysis and optimisation. Other, not so familiar keywords such as *numerical simulation of differential equations*, *non-local integro-differential equations*, *stochastic partial differential equations and differential algebraic equations* suggest that analysis also contains topics that are not dealt with in the Secondary mathematics classroom. So where do we go from here?

A lot of mathematics teachers demand of themselves, and of their students, that they understand the mathematics covered in class as completely as possible. Often, rather tricky areas of mathematics are not dealt with in secondary school as a result of this demand. The topics frequently omitted in favour of calculus might be infinity, the completeness of real numbers or the proof of the transcendence of *e* and *pi*. On the one hand, we can understand where this is coming from. After all, what teacher wants to experience a situation in which he/she is not even able to answer a question after a lengthy period of consideration because he/she lacks sufficient expert knowledge? But on the other hand, visiting these tricky topics is a risk well worth taking by teachers as it makes mathematics class more interesting and motivating. To this end, one of the authors and a colleague (J. Maaß and W. Schlöglmann) suggested reporting on mathematics-in-industry or technomathematics projects during mathematics class to the ISTRON Group more than 20 years ago (Blum 1993). At the time, the proposal met with a lot of criticism by the ISTRON-Group and it was repudiated by most colleagues. The main argument against including these tricky topics is that mathematics teachers are not supposed to deal with a topic in their mathematics class that they are not completely familiar with. However, this condition applies to hardly any models set up by experts, whether the focus is on numerical analysis, fuzzy logic, geometry or Gröbner bases. The mathematical methods used in such models are often at the level of doctoral dissertations or above. Consequently, mathematics teachers cannot be expected to cover them in class with their students. Also it seems unreasonable to apply standards to teachers that expert modellers do not meet in their own work.

At the same time, the question of why we are dealing with a certain topic in the first place often arises during mathematics class. Seen in this light, we should reassess the question above. Should we incorporate the latest research and new research results into mathematics classes as other school subjects do, or do we simply forego the opportunity?

If a mathematics teacher is not able to explain all the mathematics used in detail, this is no reason to feel ashamed. Would an Arts or language class not seem very constrained if the teacher was only able to discuss the works they created or could create? How much of its positive image does biology owe to the fact that medical successes against epidemics or cancer are discussed, without the biology teacher ever having conducted any medical research themselves? Similar reasons may also be found for other areas of instruction.

In line with the style of this book we shall end our general discussion/reasoning here and move on to what is hopefully a motivating example of a report on mathematics-in-industry compiled from the internet. We found it intriguing even though we lack the medical background required for such a research project.

6.1 MRI (Magnetic Resonance Imaging)

Let us start with a well-known medical device/process, the Magnetic Resonance Imaging (MRI) diagnostic tool. Unlike X-rays, the type of ray used in this device is considered absolutely harmless to the body. Nowadays it is possible, as a result of the imaging capability of the MRI and other diagnostic tools, to screen the human body without having to cut it open. The MRI works by sending waves of various wavelengths into the body. The resulting reflexions, attenuations, diffractions and so on are measured and the ensuing results are converted into images showing a section of the heart, the brain, the spine or other body parts. Suitable software then assembles multiple images into one total image. State-of-the-art developments even make it possible to display organs or joints in action and thus facilitate better diagnosis. It is no exaggeration to say that the advances of modern medicine are based largely on the significant progress made in the field of diagnosis.

A very extensive and professional report on the topic is available at https://en.wikipedia.org/wiki/Magnetic_resonance_imaging and we invite you to read it. From a mathematical point of view we recognise plenty of mathematics, but the fundamental and decisive progress is not attributed to mathematics. This is true of most technologies! So what has all this got to do with mathematics? How can we identify the connections with mathematics? Let us look at the History section (Sect. 5, paragraph 7) of the report. This states:

> Paul Lauterbur at Stony Brook University … developed a way to generate the first MRI images, in 2D and 3D, using gradients. In 1973, Lauterbur published the first nuclear magnetic resonance image.

Where is mathematics contained in this? It is contained in the gradient fields. What on earth is a gradient field? In mathematics, the *gradient* is a multi-variable generalisation of the derivative. While a derivative can be defined on functions of a single variable, the gradient takes the place of the derivative when dealing with functions of several variables. The gradient is a vector-valued function, as opposed to a derivative, which is scalar-valued (https://en.wikipedia.org/wiki/Gradient).

Our second example of mathematics in this field relates to the work of Axel Haase, Jens Frahm and Dieter Matthaei. In 1985, these men created the fast imaging technology FLASH that reduced, by a factor of almost 100, the time taken to produce an image via MRI technology. Of course, it is thanks to mathematics that things sped up this much! The link http://www.mpg.de/606306/pressemitteilung20100830 (which can be translated online) to a press release by the Max-Planck-Society sheds some more light onto the affair:

> **Less is More: acceleration through better image computation**
>
> For the breakthrough to measurement times, which are only a fraction of a second, several developments had to be linked successfully. The scientists used the FLASH technology again, but this time with a radial coding of the location information, which makes the MRI recordings largely unaffected by movements. In order to shorten the measuring times further, mathematics was required. "There are significantly less data than was added for the calculation of an image are usually necessary. A newly developed by us mathematical method makes it possible for us to calculate a meaningful picture of fact, incomplete data," said Frahm. In an extreme case, a comparable good image can be calculated from only five percent of the data of a normal MRI image, corresponding to a 20-fold shorter measurement time. The researchers from Göttingen have therefore accelerated the MRI measurement time by a factor of 10,000 in the mid-1980s.

Students should be able to create more than a simple presentation using this general information and everything else available on the topic on the web. We suggest you get your students to focus on a typical application of MRI for their student presentation, such as examining joints or internal organs. Lately, it has also become possible to observe these body parts in real time. This approach not only provides highly detailed images, but it also creates a link to students' everyday lives and it makes them realise how they can benefit from mathematics in the long run. This realisation will be reinforced by teachers' attempts at cooperating across school subjects, for example, biology and physics.

Here is another suggestion for you and your class. For those of you who wish to find out more about how reflections and measurements can be used to process an image of an internal object, we recommend the following experiment. Place a simple geometric object such as a book or a wastebasket, on the floor of the classroom. Ensure that the object cannot be seen by placing a cardboard sheet (represented by the circular area in Fig. 6.1) or some other material on top of it, making sure you leave enough space to roll a ball underneath. Now roll a ball towards the visually obstructed object and using reflections and measurement ask your students to try to estimate certain features of the object such as its shape, dimensions and density.

How many times did the ball have to be rolled from various directions, so that the mysterious object at the centre of the area hidden from sight could be determined to a sufficient degree of accuracy? Again we must ask, what do we actually mean by *sufficient degree of accuracy* here?

Now that we have discussed some of the mathematics underpinning MRI imaging, is this a topic that you would be prepared to use in your class? We do not know your answer at this point, but regardless we shall try to convince you of the validity of this general approach by providing you with some further examples.

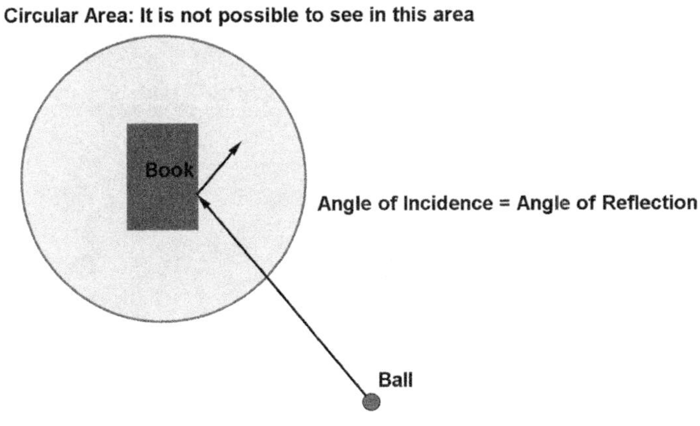

Fig. 6.1 Layout of the ball experiment

6.2 Agriculture

Agriculture has been making use of mathematics in several ways for a long time. This may be to optimise feeding, to perform business calculations or to determine how to best use seeds or fertiliser. Thanks to Professor Peter Gritzmann (TU Munich) and his colleagues, a novel and economical application of mathematics has become available in this domain. His deliberations started with a typical appropriation of agricultural land in the countryside, as shown in Fig. 6.2.

Over the course of time, a pattern of small widely scattered fields, similar to a patchwork quilt has evolved. However, every agricultural enterprise aims to have a rather large and interconnected area close to its farmhouse at its disposal. Extensive areas of land also make the use of large agricultural machines feasible. Even though it

Fig. 6.2 Conventional method of optimisation in agriculture (for colour version see http://www-m9.ma.tum.de/Projekte/LandConsolidation)

Fig. 6.3 Optimisation in agriculture employing the new method (for colour version see http://www-m9.ma.tum.de/Projekte/LandConsolidation)

is more expensive to purchase large tractors and harvesters, they allow more extensive areas to be worked, for example harvested or ploughed, in the same amount of time it would take to work smaller pieces of land. If a thunderstorm looms and a lot of grain needs to be harvested, time is an important factor. Obviously, shorter transport routes, close to one's farmhouse, are preferable to longer drives to a far-off field.

It would definitely make sense to all the parties involved, and also from an economical point of view, to merge and swap fields so that everyone can better economise. A traditional approach to improve the situation would be land consolidation. This however is a time-consuming, governmental procedure that does not always meet with success. Figure 6.2 shows the conventional situation after land consolidation. This situation motivated the project team at the University to ask if there was a better way to do this. The team then proceeded to develop a rather costly mathematical method (model) to assess the value of individual properties and calculate opportunities for a fair exchange based on factors such as soil quality, yields, distance to the farmhouse

Fig. 6.4 Conventional method for optimisation in agriculture (Example 2)

and last but not least the agricultural subsidies paid by the EU for these properties. Figure 6.3 shows the results.

In this case, a picture (or better, a PowerPoint slide comparing both pictures) really is worth a thousand words. Obviously, they were able to optimise the arrangement of landholdings, as the blue and green fields have now been merged. If students realise the economic benefit resulting from this rearrangement, they have understood an example of mathematics applied to real-life without having looked into the optimisation technique in closer detail.

Here is another example of mathematical modelling that is even more impressive. In this example we have an agricultural area with 7 farmers and 419 plots (Fig. 6.4).

Figure 6.5 shows the results after optimisation.

Fig. 6.5 Optimisation in agriculture employing the new method (Example 2) (images courtesy of Prof. P. Gritzmann (TU Munich))

6.3 Transport and Logistics

Anyone, who sees convoys of trucks on the motorway or cargo trains that seem endless, can appreciate that goods need to be moved from A to B, on time, to ensure supplies are available for production, even if they are not familiar with the industry practice known as just-in-time production. Another problem, the travelling salesman problem, which deals with the question of the perfect route for a salesman, is well-known even outside of mathematics. The salesman must arrive at several possible points of sale as quickly as possible. Trucks that service several branches of a retail chain also face similar problems. Another mathematical problem is how cars, consisting of thousands of parts produced in different places, should best be assembled. Typically, the parts are manufactured in many different locations. One subcontractor produces tachometers, another one the rims for the tyres, another one the control systems, and yet another one the brakes. Furthermore, the engines are fabricated in a different plant. All of this should be coordinated in a way so that all parts required for assembling a car should arrive at the assembly plant on time without having to store them anywhere in the meantime (storage costs more money and incurs more expenses for transportation). Also it is imperative not to keep the assembly line waiting for parts—that too costs a lot of money. It goes without saying that planning has to be accurate and that mathematics is required for planning and optimisation in this case. The Research Institute for Symbolic Computation (RISC-Institute) is an example of a research centre which, since its establishment in 1987, has dealt with these types of problems in detail (see http://www.risc.jku.at).

These problems are really easy to understand and simulate. Even if you only use a few locations and demands on transportation, you can comprehend and understand these problems. We suggest using toy cars and Lego bricks that can be observed and analysed. Using this approach it is easy to see how complexity increases for our travelling salesman as more and more points of sale are added. Transport and logistics are key activities in business. Should you wish to deal with these topics in class and look for particularly interesting and up-to-date problems, we recommend the Journal on Transportation and Logistics (http://www.springer.com/business+%26+management/operations+research/journal/13676).

The mathematics required here is still within the realms of mathematics for secondary schools (linear optimisation, graph theory, and network analysis) and the business facts can be easily understood without having done business studies. Nevertheless, you will need to discuss your modelling assumptions in detail, that is, what should be modelled and how exactly can we do this.

6.4 Distribution of Heat When Re-entering the Earth's Atmosphere

Those of you interested in human explorations of space will have noticed in documentaries or Sci-Fi movies that space shuttles experience extreme frictional heat when they re-enter the atmosphere, despite the very thin atmospheric layers. Space shuttles that are re-entering too fast or attempting re-entry at a wrong angle are even in danger of burning up. Whoever wants to build a space shuttle that is to be re-used needs to consider carefully how it should be built so that it will not suffer any heat damage or be destroyed on its return to Earth. The excessive cost and landing conditions make tests virtually impossible. Therefore, prior considerations and simulations are essential. But how can flight landings be simulated? By mathematical modelling of course! NASA's Ames research centre has investigated this issue for many years and offers some insights, resources and possible topics for student presentations in relation to the mathematics of space exploration (see: https://www.nasa.gov/ames).

Additionally, if you take a closer look at the extensive list of publications by the task force 'Technomathematik' (http://www.mathematik.uni-kl.de/forschung/), many more topics suitable for student presentations on successful applications of mathematics in technology will present themselves.

6.5 Identifying Topics Suitable for Student Presentations

Another source for motivating student presentations on real-life applications of mathematics and on successful modelling of reality is information provided by the Max Plank Society (http://www.mpg.de/). For those from a German speaking country there is also a new portal offering short video clips on topics taken from (popular) science http://maxplanckcinema.tumblr.com/. At first glance, videos like these may appear to be only about physics, geophysics or biology, but not mathematics. However, as soon as we click play, we hear typical keywords that tell us mathematics is being applied here, for example computer simulation, calculation, modelling, optimisation, relativity and so on.

If we randomly open the website of one of the many Max Plank Institutes and search for Press Releases, the result is similar. We found the following press release on 'Controlling medical nano-deliveries' (https://www.mpg.de/606935/pressemitteilung201006083?%20filter_order=L&research_topic; Note: You can use Google Translate to get this press release in English).

This article might not, at first glance, appear to involve mathematics, but the subtitle highlights that some mathematics is involved here: *A model that allows us to predict how efficiently nanoparticles attach to the surfaces of tumour cells.* Clearly a mathematical model is involved here. A biological process is being modelled, which is very interesting from a medical perspective.

6.6 History as a Source of Modelling Problems

Mathematics can be used to solve problems related to nanotechnology, nuclear power stations, cosmology, genetic engineering, chemical engineering, mechatronics, economics and many other contemporary fields of knowledge. This may be very intriguing, but it can also act as a deterrent. Negative motivation may be triggered as new technologies are novel and unfamiliar, but also some may pose possible risks to humanity and the environment, as has become apparent in the case of nuclear energy. Therefore, another option would be looking back in history in our quest to find examples of mathematical models. Historically the evolution of mathematical ideas and theories involved many great mathematicians encountering setbacks and problems. These problems and setbacks are often not shared with students today despite their potential to humanise and arouse interest in the subject of mathematics. To help in this regard, the society of didactics of mathematics has set up a taskforce for the history of mathematics. Its main aim is to make the history of this science more accessible to mathematics educators.

Here is just one example for you. Biermann has conducted a study on the 18th century mathematician J. H. Lambert (1728–1777), a member of the Academy of Sciences in Berlin and his work, in which he describes Lambert's activities as follows:

> The viability of technological inventions had to be assessed, optical and mechanical instruments held back by customs had to be examined, salt production had to be optimised, construction plans for a plant producing sulphuric acid had to be reviewed, a machine had to be assessed, a watering machine at the botanical gardens had to be tested, a well builder's request for an advance payment had to be scrutinized, the curriculum for philosophy at a university had to be reviewed and so on. His findings were then summarized in a written report … Please find below a few examples to illustrate the manifold topics Lambert had to deal with: Inventing a spinning frame, improving astronomical tools, a paper on gnomonics, a paper by Baedov on education, proposals on successful fire-fighting, a publication on the use of various soils, technical designs, a memorandum on a potato mill, a publication on the use of mineral waters, another one on hydraulic engines, proposals on how to solve the problem of squaring the circle, remarks on Kästner's mine surveying science in Göttingen, a paper on the magnetic properties of the ether, a description of Wilson's experiments with light and colour in London, new theories on the planets and comets, a treaty on the universal language translated from Hungarian, the construction of a sick-bed [his assessment of it was so important to Lambert that he had it printed in the publications of the academy (Biermann 1979, p. 1187)].

Such texts may be used for research tasks as follows: What has Lambert (or somebody else) published on potato mills? What has this got to do with mathematics? If this research bears fruit, a new project might even develop: Let us calculate and build such a mill ourselves! Regardless of whether or not your students can put this mill to use in their everyday lives, our research on 'Adults and Mathematics' proves that people will look back on this special mathematics education and remember it in a positive light decades later. Unfortunately, very few adults have any such positive memories of their mathematics education to look back on.

We end the chapter by encouraging teachers to use these kinds of expert problems to develop their own understanding and to involve their students in motivating and engaging classroom units. In our view, this practice equips students with a more positive experience of mathematics and a deeper appreciation of the subject, as a basis for lifelong learning.

References

Biermann, K. R. (1979). J.H. Lambert und die Berliner Akademie der Wissenschaften. In Colloque Intern. Jean-Henri Lambert (1728–1777). Paris.

Maaß, J., & Schlöglmann, W. (1993). Der Stoßofen - Ein Beispiel für Industriemathematik als Unterrichtsthema. In W. Blum (Hrsg.), *Anwendungen und Modellbildung im Mathematikunterricht* (pp. 74–84). Hildesheim: Verlag Franzbecker.

Chapter 7
Further Tips for Teachers Who Want to Implement Applicable Mathematics Education

In this chapter, we ask you, the teachers, and your students to reflect on your learning journey as you progressed through the book. We invite learners to follow a structured approach based on the layout of the book and to question key facets of their journey. An approach to this practical, personal evaluation is presented in the first part of the chapter and deals with suggestions for the preparation of applicable mathematics lessons, followed by a section on finding new problems, and concludes with templates for appraising students' progress. Teachers need to know where this approach leads in terms of students' future learning and their studies and even employment.

We believe it is time for a preliminary review: Which topics have we dealt with so far? What is still missing? At this point you are also invited to conduct your own preliminary assessment. A preliminary assessment is also the key to successful learning for your students.

Have you been following our reasoning so far?
Have you gained some insights for your own classroom practice and have you experimented with it already?
How can you assess what your students have learned about modelling, without necessarily doing more modelling projects?

Our preliminary review will be based on the structure of the book. The following questions guided the structure of this book and the answers will therefore be central to our preliminary review:

- What is applicable mathematics education?
- Why should I engage in applicable mathematics education as a mathematics teacher?
- How do I go about doing that?
- What options do I have, if my students are not able to, or do not want to, engage in applicable mathematics education?

J. Maaß et al., *Mathematical Modelling for Teachers*, Springer Texts in Education, https://doi.org/10.1007/978-3-030-00431-6_7

- How do I get started?
- What topics could I focus on to develop more extensive classroom units?
- Where can I find more examples?

In Chap. 1, we used the interactive cell-phone problem to facilitate the process. As it was based on empirical research, it was expected to boost motivation and hence support our arguments in favour of applicable mathematics education. At the same time we referred to the various government demands in the shape of curricula, catalogues of competencies and other related demands on mathematics education and suggested that these could be reconciled with the proposed approach to applicable mathematics education.

Those teachers who are already convinced and motivated, should be in a position to begin this process themselves but bear in mind that small steps are always easier to make than large strides! For this reason, we started out with small steps in this book, but along the way they got bigger, and finally we got a glimpse of how experts model.

If you are prepared, at least to some extent, and able to deliver applicable mathematics education, you will realise, like many teachers before you, that this type of instruction provides a special sense of achievement. You might also agree that preparing for such classes requires additional effort. The more you are willing to diverge from the path predefined by your textbook and other available materials, the more time you will need to invest in your preparation. Moreover, if you are not used to this style of teaching, you may not benefit as much from your routine in preparing your classes as you did for conventional mathematics classes.

This raises another question: Should we feel guilty about proposing a teaching style that requires more effort than 'conventional' classes? Our answer to this question is 'No!' Why should we? You will be rewarded by a sense of achievement during class and with positive feedback from your students. On a positive note, we believe this book can help you limit your additional effort by providing some suggestions that will be explained in more detail in the following section. As we do not know how you have been working and preparing your classes to date, we might suggest things you have already been doing for a long time. If that is the case, you should view our suggestions merely as an acknowledgement of your innovative approach to teaching. On the other hand, our aim in providing these suggestions is not to claim that we know more than you do regarding how you work, instead, our goal is to support teachers who would like to adopt a more applied teaching style, but initially feel overwhelmed by it.

7.1 Suggestions for Preparing Applicable Mathematics Lessons

Suggestion 1: The extent or completeness of our preparations
We know of many teachers who believe in determining everything in advance of class. They decide on the content to be covered during a particular period and identify which solutions are admissible. This sounds more time-consuming than it actually is, particularly when the goal of the lesson is to solely work through problems selected from a textbook that also contains the solutions. In contrast, it takes much more effort to prepare an applicable mathematics lesson, which leaves many of the classroom decisions up to students, for example whether or not they want to perform another modelling cycle with new assumptions or additional data. In terms of teacher preparation, if a teacher attempts to anticipate all possible questions and ideas relating to the model, and aims to calculate all results in advance the workload increases immensely. In our view it might make more sense to change our approach to preparing our lessons. What do we mean by that? We suggest that teachers instead focus on preliminary considerations and then work together with the class on the model. This alters the role of the teacher. He/she is no longer responsible for all matters pertaining to the content and the organisation of the project and does not have to answer all questions correctly and in full as soon as students pose them. Instead the teacher focuses on assisting the students in coming up with questions, answering them and finding the required data themselves.

The first thing we need to consider when preparing applicable mathematics classes is whether a certain problem taken from reality suits this particular class? Two things are crucial here; the topic and the curriculum. The mathematics required for our model needs to match the curriculum of this class. It is a hard task to assess how motivating any real life topic actually is. In the first chapter, we made a decision to start the ball rolling by assuming that most students use a mobile phone and prefer to pay as little as possible for using it. Therefore, we concluded that 'mobile phone rates' is a topic that they can relate to and find motivating. It is not hard to convince them how this topic affects them as individuals responsible for their own actions. Do these assumptions also apply to electricity rates? We are not quite sure. We can definitely relate to this topic, as we have to pay electricity bills regularly. Will we succeed in convincing our students that they too will soon have to pay electricity bills themselves and therefore the topic is highly relevant to them? Or do we try to motivate them by asking them to give advice to somebody who is looking for a suitable electricity rate? Inventing a little story about Grandma Miller from next door might prove useful here. Maybe she has received an offer from an electricity provider but does not understand it. What advice can the class provide? Would it be better for her to switch to another provider?

When preparing applicable mathematics lessons, obviously it is very important for the teacher to incorporate the curriculum and consider what mathematical methods are required for dealing with the topic chosen. Here are some extreme examples: recalculating a receipt or a restaurant bill requires adding up figures with two digits after the decimal point; whereas in case of rates we need linear functions and equations; and optimising winter tyre treads for cars calls for mathematics way beyond school levels. It is not difficult to identify the appropriate school level that each these extreme cases are suitable for. It is much harder to find the appropriate level if the mathematical content of two possible projects is similar, or the curriculum assigns the mathematical methods required for modelling to different school levels. It is much easier to make this decision, or do the preparatory work, if reports on dealing with the selected topic in mathematics class are already available. This is what the next suggestion refers to.

Suggestion 2: Make use of existing resources
In general, the motto 'Reading fosters knowledge' holds true in the case of applicable mathematics education. The reported experience of other mathematics teachers and other publications can prove particularly useful to mathematics teachers who plan to teach applicable mathematics classes. We will now offer a number of suggestions regarding websites and journal articles that may help you to source ideas for applicable mathematics education in your own classroom. We expect active teachers will expand/develop their own library of resources in this way as they continue to engage with applicable mathematics in the classroom.

Useful Websites and Articles

http://www.nctm.org/publications/mathematics-teacher/

This website directs you to the magazine 'Mathematics Teacher' produced by the National Council of Teachers of Mathematics (NCTM) in the United States. The highly reputable, monthly magazine offers suggestions and ideas for teaching mathematics.

http://www.comap.com/Free/

This is the website for the Consortium for Mathematics and Its Applications (COMAP). This U.S. based organisation run an annual High School Mathematical Contest in Modelling and this website offers a wide selection of problem scenarios used in the competition.

http://www.mathalicious.com/

This website is designed to provide American teachers with lesson plans and activities that challenge students to think about the mathematics that appears in the world around them. They offer ideas for modelling problems that are relevant to middle and high school students. While this is a subscription based website there is an offer of a thirty day trial.

https://www.middleweb.com/5003/real-world-stem-problems/

This website offers suggestions for teachers of all middle grade students in the U.S. While this website is designed for teachers of middle grades the scenarios offered can be adapted and extended by resourceful teachers for all grades. The site discusses a number of STEM related problems that could be used in the mathematics classroom and also contains a section that directs you to mathematics specific problems.

http://epistem.ie/resources/teaching-learning-resources/real-world-applications-secondary/

EPI•STEM, the national centre for STEM education in Ireland, in collaboration with the Engineers Ireland STEPS programme, developed a series of information leaflets and worksheets which show the importance of mathematics in the real world. These worksheets, which are available from this website and were developed for secondary school students, outline realistic problems encountered in a number of different industries and show how mathematics can be used to solve such problems.

https://www.ncetm.org.uk/

This is the website of the National Centre for Excellence in the Teaching of Mathematics (NCETM), based in the U.K., and it offers resources and guidance for secondary mathematics teachers. It also contains a number of reports, such as the *Learning Outside the Classroom* report which offers a rationale for engaging in applicable mathematics as well as ideas and resources.

https://link.springer.com/article/10.1007/s10857-007-9070-8

This is a link to a journal article by Gainsburg (2008) entitled *Real-World Connections in Secondary Mathematics Teaching*. This is an empirical study of how teachers engage with applicable mathematics in U.S. classrooms.

At this point we would like to remind you that this is not an exhaustive list of sources. Furthermore, internet resources are evolving all the time and so teachers may need to factor in time to keep up. When looking for further resources, a little browsing (i.e. title and abstract) is usually enough to find some proposals for your class and the course content you are currently dealing with. It can save a lot of time and energy to select a lesson proposal for your class which has already been published and then adapt it to your own needs, as opposed to developing something from scratch yourself. 'Adapting' may sound like extra work, but it does not necessarily have to be. Maybe the students can do this work! For example, mobile phone tariffs constantly change. The task becomes a lot less motivating if the lesson contains tariffs from years ago just because the proposal dates back to that year. Teachers do not necessarily have to research current mobile phone tariffs and incorporate them into the lesson. Instead, this task should be viewed as a student activity as part of the actual project.

Suggestion 3: Cooperate with other teachers

It might make sense to cooperate with mathematics teachers from other schools or teachers of other subjects in your own school. On the one hand, our experience (and feedback from many schools) suggests that it is often difficult or even uncommon to cooperate with teachers in your own school. On the other hand, we often hear teachers report enthusiastically about successful collaborations. We do not know anything about the particular state of affairs at your school and therefore we cannot provide you with any personal advice. However, looking at the positive feedback we received so far in relation to collaborations, we were able to identify one common feature: start your collaboration with a small project and as the joint interaction develops, the quality of the cooperation will also improve.

The feedback we received on collaborations was predominantly positive and our experience confirms that this response is widespread. Mathematics teachers who were feeling isolated in their own school, as they may have perceived themselves as the only ones who wanted to improve mathematics education, really enjoyed meeting mathematics teachers from other schools with similar experiences. So, how do you go about finding such colleagues? Here are two suggestions: you can find like-minded colleagues by attending continuing education events for teachers (often highly motivated colleagues from the same area meet there) or alternatively you could participate in local task forces or committees.

7.2 Finding New Problems

In the past, blank areas on the global map have prompted many people to make great efforts and venture into the unknown. However, discoveries less significant than the discovery of the source of the Nile, can also have their charm. So, what new discoveries can you as a teacher or your students make in mathematics education in school? All definitions and theorems dealt with in mathematics education were discovered and proven a long time ago. This leads us to a significant advantage regarding applicable mathematics education; in this field new discoveries become possible for you and your students. Each model of a real situation is a new creation, the discovery or construction of something previously unknown. Even if somebody else has modelled mobile phone tariffs in a similar fashion before, we are pretty convinced that there will be aspects of your model that other models do not contain or aspects that are presented in a different manner. Thus, independent modelling opens up new horizons for mathematics education. It is not about comprehending preconceived knowledge or practicing how to handle well-known algorithms, but is concerned with conceiving and shaping our world in a creative way. Over the course of its history, this has always been an integral part of mathematics. However, as mathematical research has long surpassed the level of school mathematics, this dimension usually no longer features. This is one of the reasons why mathematics at school appears rather rigid, stagnant and not very appealing, but unjustly so.

Once your interest has been awakened to experience mathematics in a different way, to model something yourself, the following question arises: How do we go about identifying a suitable topic for our class? As with our previous suggestions, you might already be aware of potential topics. If you are not, a possible suggestion is to read newspapers and magazines or internet reports, scanning them for up-to-date and motivating topics for your lessons. Plenty of reports contain statistics/graphs or others ways of displaying statistical or mathematical models, that support the authors' argument. Often these graphs prompt discerning questions, as axes might be shortened, only certain parts of the data are analysed and so on. As these reports frequently advertise or protest against decisions of personal, political or economic importance, it is crucial for students to learn to see them in the 'right' light. If mathematical methods enable us to identify which part of the message is enhanced by choosing a certain type of graph, it makes it much easier for us to comprehend the text. In this way, mathematics directly helps foster critical minds. For further reading on these ideas, we suggest the work of Garfield (2002).

We should be careful to recognise that all types of media, including newspapers and media broadcasts, use graphics in selective ways. The point is that these graphics are not always a neutral representation of data but are meaningful and suggestive pictures that are used to "prove" what the author wants to say. Nowadays, we are regularly confronted with the new phenomenon of *fake news,* which invariably makes use of statistics in misleading ways (https://www.forbes.com/sites/kalevleetaru/2017/02/02/lies-damned-lies-and-statistics-how-bad-statistics-are-feeding-fake-news/#c1866cd50ca1). In Chap. 4 we saw how data can be manipulated visually to support a certain argument. For example, if commentators want to express differences they may use columns instead of lines (the brain sees a three-dimensional object and not just the difference between two dots). Other well-known tricks include cutting graph axis, changing the scale and so on. Many more examples and explanations of the use of misleading statistics are provided on the following website http://faculty.atu.edu/mfinan/2043/section31.pdf. If teachers wish to understand how statistics are used legitimately, or in a misleading way, they may use existing data relating to topical issues such as climate change, Brexit, or CO_2 emissions. The following websites are good sources of material that could be used to further explore this idea:

Climate Change: https://climate.nasa.gov/evidence/
http://edition.cnn.com/2016/12/07/africa/sudan-climate-change/index.html
Brexit: http://www.telegraph.co.uk/news/0/eu-referendum-claims-won-brexit-fact-checked/
https://opentoexport.com/blog/ioe-share-findings-from-their-brexit-survey/
CO_2 Emissions:
https://www.theguardian.com/environment/2016/nov/14/fossil-fuel-co2-emissions-nearly-stable-for-third-year-in-row

In addition to sourcing teaching materials from print and digital media, we will now suggest two additional approaches for finding suitable classroom problems. Our first suggestion is to concentrate on a type of problem called a *Fermi-Problem.* An exemplar Fermi problem asks us to consider *how many piano tuners are there in Chicago?* A typical solution to this problem involves multiplying a series of estimates that yield the correct answer if the estimates are correct. For example, we might make the following assumptions:

There are approximately 9,000,000 people living in Chicago.
On average, there are two people in each household in Chicago.
Roughly one household in twenty has a piano that is regularly tuned.
Pianos that are tuned regularly are tuned on average about once a year.
It takes a piano tuner about 2 h to tune a piano, including travel time.
Each piano tuner works 8 h a day, 5 days a week, and 50 weeks a year.

From these assumptions, we can compute that the number of piano tunings in a single year in Chicago is:

(9,000,000 persons in Chicago) ÷ (2 people per household) × (1 piano per 20 households) × (1 piano tuning per piano per year) = 225,000 piano tunings per year in Chicago.

Similarly, we can calculate the number of tunings that the average piano tuner performs per year to be:

(50 weeks per year) × (5 days per week) × (8 h per day) ÷ (2 h to tune a piano) = 1000 piano tunings per year.

Dividing these figures gives:

(225,000 piano tunings per year in Chicago) ÷ (1000 piano tunings per year per piano tuner) = 225 piano tuners in Chicago.

The actual number of piano tuners in Chicago is about 290 (https://en.wikipedia.org/wiki/Fermi_problem).

We think this is a motivating example. Fermi started with several assumptions and ended with an astonishingly close estimate to the true answer. If you wish to train your students to become excellent estimators, you can find a lot of resources on the internet, for example this one: http://www.edgalaxy.com/journal/2012/5/29/an-excellent-collection-of-fermi-problems-for-your-class.html.

Our second suggestion is to focus on the concept of *Maths Eyes* (http://www.haveyougotmathseyes.com). This Irish project has opened our eyes to a rather different but very productive approach to applicable mathematics education. This idea has been taken up in Linz and Koblenz where they have held their own competitions (http://www.jku.at/idm/content/e83438/e209929 and http://matheyes.uni-koblenz.de/de/). At this point we resist the temptation to fill this book with pictures of competition entries. However, in the next chapter we will present you with more details regarding this initiative and some thoughts on its background. At this point, we will simply show you (Fig. 7.1) one of a number of examples submitted by students as part of the Irish competition.

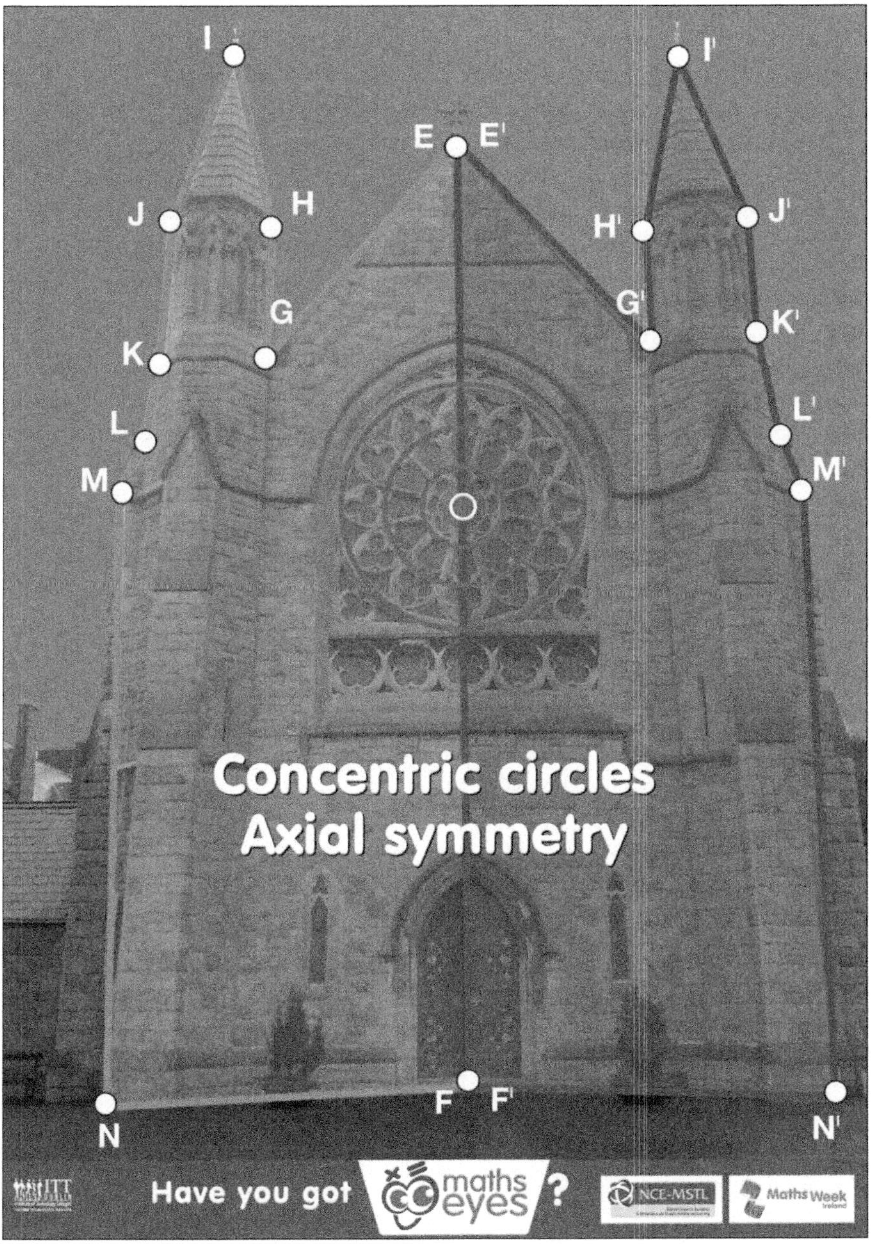

Fig. 7.1 Maths eyes competition winner (see colour image on: http://www.haveyougotmathseyes.com/wp-content/gallery/shortlist-2013-2/A3-Portrait-Template-page2.jpg)

7.3 Preliminary Assessment of Your Students' Progress

In our view it makes sense to stop along the way, look back and consider how things have progressed and think about how we should proceed from here. Therefore, in this chapter we not only recommend that you reflect on your learning journey but that your students reflect too. It is always worthwhile to ask a number of questions, for example, what have we learned about this topic so far? What would we like to know more about? What would we like to do in the future?

Our main hope and expectation for applicable mathematics education is a permanent increase in student competencies. This does not mean that positive short-term learning outcomes do not occur or cannot be assessed—on the contrary! Obviously, independent mathematical modelling skills are best appreciated if you actually do some modelling yourself. In our quest for less time-consuming assessment methods and ways of doing a preliminary assessment, we have come across the following proposal that has been developed in the course of the LEMAMOP (Learning Opportunities for Mathematical Reasoning, Modelling and Problem Solving) project. It has also been presented during an ISTRON conference in November 2014 in Koblenz. This approach explicitly avoids recalculating or assessing a calculation that has been presented, and focuses specifically on modelling assumptions. Here is the first example.

Example 1: Traffic Light Problem
Simon has to pass three sets of traffic lights on his way to school. They display green and red lights for the following durations:

Traffic Light 1: 30 s green, 50 s red
Traffic Light 2: 20 s green, 30 s red
Traffic Light 3: 15 s green, 30 s red

If you wish to calculate the probabilities for different events, for example Simon encounters no red lights; all green lights or 2 red lights and 1 green light, the situation is often modelled using a tree diagram (see Fig. 7.2). However, this is only one of a number of possible solutions.

Task: Please explain which aspects of the real situation are represented well in this diagram. Is it possible that this diagram is adequate to analyse the situation even if it is not optimal?

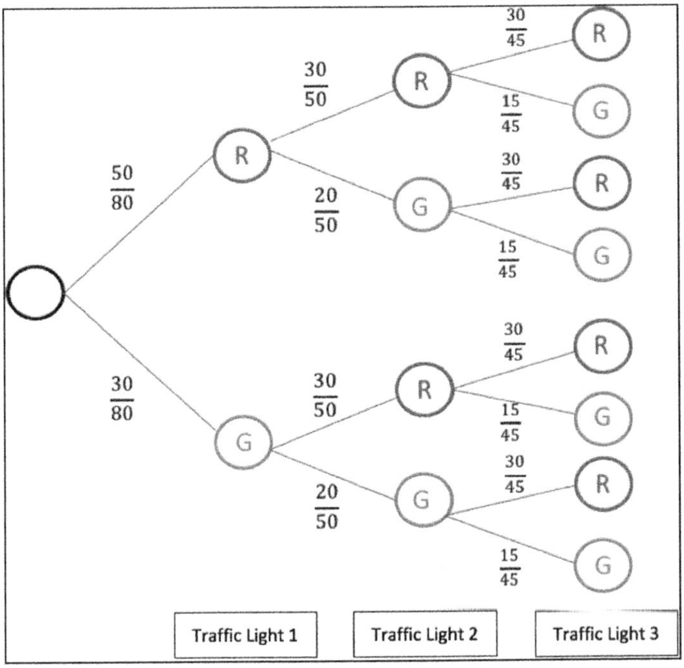

Fig. 7.2 Problem for assessing modelling (slides by Bruder and Hinrichs, p. 22, kindly provided by Prof. Dr. R. Bruder)

In our view, this example fulfils the vital requirement of relating mathematics to reality, which makes permanent learning outcomes in mathematics education possible. Students who have frequently come across such tasks in mathematics class will no longer ask why they need to study mathematics—now it is apparent how mathematics contributes to the educational goal of 'learning for life'.

Having looked at one possible solution, now it is your turn:

How would you go about solving this problem?
Which solution would attract full marks?
For students who use a different model to solve the problem, than that offered in Fig. 7.2, how would you grade it?

We will now look at a second example. The following example is often presented as a joke on poor mathematics education in which the problem features too prominently: How many painters are needed for painting a flat in a second?

Example 2: The Craftsman's Problem
Three craftsmen need 2 days to paint all the walls of a new apartment. How many days would 10 craftsmen would need?

Your student proceeds as follows:

Number of craftsmen	Days to work
3	2
1	6
10	0.6

Your student reaches the conclusion that it takes 10 craftsmen approximately 5 h to complete the job.

How realistic is this model?
What would your answer be and what would you evaluate as a good answer?

Throughout the book, we have asked you to engage with modelling by guiding you through a range of modelling processes in a formative manner. We now suggest that you as a teacher adopt a similar approach with your students by concentrating on the modelling processes using a number of different techniques such as questioning, observation, presentations or feedback. Formative assessment focusses on student learning. Formative assessment guidelines are readily available in numerous sources (http://www.siam.org/reports/gaimme.php). In addition to providing you with formative assessment guidelines, this source also discusses summative assessment rubrics. The summative assessment rubric focuses on grading/marking. The summative approaches to assessing modelling are generally divided between two aspects, namely a holistic approach concentrating on the modelling processes and a component approach based on the different components of the modelling cycle. These components are outlined in many sources already listed in this book, for example understanding the problem, formulating the problem in mathematical terms, defining the mathematical model, solving the mathematical model, validating the solution, interpreting the solution and presenting the results.

References

Gainsburg, J. (2008). Real-world connections in secondary mathematics teaching. *Journal of Mathematics Teacher Education, 11*(3), 199–219.

Garfield, J. (2002). The challenge of developing statistical reasoning. *Journal of Statistics Education, 10*(3), 58–69.

Chapter 8
Empirical Findings on Modelling in Mathematics Education

Empirical findings related to studies in English-language countries are widely disseminated but the same is not true for studies reported in languages other than English, for example studies reported in German or Spanish. It seems sensible to include in a book like this some reference to, and discussion of, examples of good practice in other non-English speaking countries and to bring this to the attention of teachers. We are very well served on the writing team by the leading author (JM) concerning access and understanding of German-language studies and sources to practices, materials and developments. This chapter also offers insights and results from translated German language studies. In this way we make selected studies accessible to English language readers. In particular, we believe that teachers and teacher educators will find this chapter interesting.

It would make life much easier for us, as mathematics teachers, if we knew in advance of trying something new, such as reality-based mathematics, that it would work out and result in increases in student motivation, understanding, and/or enjoyment of school mathematics. However, it is understood that forecasts of this nature are uncertain because the predicted events or outcomes lie in the future. Consequently, neither we nor anybody else can assure you that your first attempt, or subsequent lessons, will be a success. However, we do believe we can help improve your lessons and that is what this book is all about.

In this context, we need to caution you to beware of 'recipes' for the perfect mathematics class that are supposed to work anywhere and at any given time. This is simply not possible because learners and teachers are individuals who engage in different settings in a certain context equipped with diverse experiences and expectations. A recipe, on the other hand, requires that the ingredients be identical, no matter who the cook is, or where it is being cooked. Continuing this analogy, 20 g of salt will always be 20 g of salt, but 20 students in one context is not identical to 20 students in another context. Similarly, one teacher is not identical to another! If you try the same recipe with the same ingredients time and again, the result is likely to improve. But what happens if you deliver a lesson using the same content and methods to the same group of students time and again? Such lessons are unlikely to be well received by students, or rewarding for teachers.

J. Maaß et al., *Mathematical Modelling for Teachers*, Springer Texts in Education, https://doi.org/10.1007/978-3-030-00431-6_8

Proposals for course changes can only work if you, as the expert on your own teaching, embrace suggestions and put them into practice in a way suitable for your class. As nobody else possesses such in-depth knowledge of your mathematics classes, successful implementation of such changes is in your hands. There are two main routes that lead to changes in your classroom. You are currently taking one of them by reading this book—you are acquiring information on how to engage with reality-based mathematics instruction. You must then decide if this information is useful for your mathematics classes, and if the answer is 'yes', work out how you can implement it. The second option is to engage in cooperation or mentoring at your school that is implemented as a consensual process: you agree on teacher observation sessions with a colleague (or an external advisor), or on team teaching, and afterwards discuss everyone's impression of your mathematics class. We will elaborate further on these options later on.

We encourage mathematics teachers everywhere to avail of the opportunities for improving their practice through CPD courses and conferences, together with subject associations (mathematics) and networks in every country including related websites. We identify some of those events and opportunities that are particularly well-aligned with the goals of this book and refer you to them throughout. They include the ICTMA biennial international conferences that produce a constant output of high quality material and publications on modelling in mathematics including school mathematics (http://www.ictma15.edu.au/related.html) and especially the MUED conferences for German-language speakers.

Even though we cannot guarantee your success and provide perfect 'recipes' for mathematics education, we would like to encourage you to follow the path to hands-on, reality-based mathematics education. It has worked well for other mathematics teachers before you. We offer some practical advice to assist you in that regard. The simplest approach to finding out more about the experiences other mathematics teachers had with such mathematics education is to talk to them. Teacher education courses and conferences provide a good opportunity for that, as colleagues that you meet there may already have gained experience with such an approach to teaching. Those of you who are making use of such an exchange of experiences will soon realise how diverse and individual these experiences are. With a bit of luck a colleague who has made such an attempt in a similar school to yours will still remember key points and share them with you. Moreover, personal contact makes it possible for you to ask questions or request for more details. Something that has not been elaborated upon in a publication might be of particular interest to you, and colleagues will gladly report it when asked. After all, it is much easier to provide information during a conversation than trying to elicit it in an email to authors at a later date.

You might have gotten the impression that such personal experiences are quite subjective, or that teachers can embellish their accomplishments or omit difficulties in their reports. This does indeed happen and in order to maintain balance, we have inserted a small chapter here in which we report some of the results of international empirical studies. Before going into detail on this, we would like to emphasize that most empirical studies are in favour of teaching mathematics in a hands-on, reality-based manner.

8.1 Empirical Research and Modelling

Traditional mathematics teacher training rarely involves any research on the didactics of mathematics or mathematics pedagogy on the part of the student, unless a written paper or a dissertation or thesis at the end of the studies contains such research. In mathematics teacher training, modules/classes on research methods tailored to the didactics of mathematics/mathematics pedagogy are not widespread and remain rather rare. This is why we will start with some remarks on empirical research on mathematics teaching methods.

The first and fundamental question deals with the purpose: Why is empirical research on mathematics teaching methods conducted in the first place? What are your thoughts on that? Our reasoning goes as follows. We consider such research very valuable, as it opens up a path to obtain feedback on mathematics-related teaching methods and suggestions. It also makes it possible to get a broader picture of the reality of teaching mathematics that is supplementary to personal experiences and individual conversations. Does this not happen as a matter of course? Actually no, so for this discussion let us look back on the evolution of mathematics education for some context.

In the following paragraphs the authors develop a broad-brush description of mathematics education as it evolved across the globe in the preceding decades. The narrative focuses on big ideas that underpin changes and can be seen to influence mathematics curricula and teaching practices. As a backdrop to these paragraphs we offer a small number of relevant observations. Historically, national governments espoused ideals for education such as: (1) citizenship and democracy; (2) training people for participation in society and the economy; (3) serving the needs of individuals for advancement. Nowadays, we can see that very often governments subscribe to a mix of these ideals as they work through their education and economic policies. Allied with this situation is the trend towards the convergence of education, economic and competiveness policies in many countries, which affects mathematics education directly because goals for education include producing mathematically literate workers. Mathematics is also seen as an underpinning discipline for STEM education, which is a major element in governments' economic strategies.

Broadly speaking, historically there has been a shift away from content-oriented mathematics curricula and teaching (traditional mathematics) to process-oriented school mathematics (reform mathematics). This worldwide trend is particularly evident since the year 2000, and is driven by a general dissatisfaction with the status quo in mathematics education. This was fuelled by the results of large-scale international comparative studies in education that focussed on mathematics and science. Globally, the Programme for International Student Assessment (PISA) and the Third International Mathematics and Science Study (TIMSS) are examples of international studies that have had a significant impact on mathematics education. The concern for process-oriented mathematics has led to a new understanding of mathematical competence that encompasses facts, skills, procedures and concepts *and* mathematical thinking, problem solving and a positive disposition towards mathematics. It is fair to say that mathematics curriculum reform is informed at policy level by research in mathematics education, but teaching practices at classroom level in many instances still do not reflect current research knowledge in mathematics education. Problem solving and mathematical modelling in applicable mathematics are consistent with the new understanding of mathematical competence and hence our interest in this approach.

For a long time, in German-language countries, research on mathematics education focused on certain areas of the curriculum, for example geometry or analysis was looked at from a certain angle in specific detail and written about in a way the authors considered to be more suitable for teaching than previous publications. Here, the works of Euclid serve as a famous example of such an attempt, as it has helped people gain a better understanding of mathematics for more than 2000 years. Didactics of mathematics have recently reached their climax with writings on new mathematics or modern mathematics, which attempt to organize mathematics in schools according to the principles of set theory (using the principles of the works of Bourbaki to be precise). We do not wish to elaborate here why these efforts failed, but we would like to direct your attention to the fact that this reform movement was not supported by any empirical research. The onset of empirical research in mathematics education in German-language countries only took place afterwards. It was probably also caused by a re-orientation of the didactics of mathematics due to the influence of foreign pedagogical theories and methods.

As is customary, we make a distinction here between quantitative and qualitative research. PISA is a typical, very complex and expensive example of a quantitative study. A large number of students are tested and surveyed when they complete a questionnaire and they are asked for their opinions. The results (for example test scores or checkboxes on questionnaires) are statistically evaluated and then interpreted. Usually it is fairly easy to link what was checked or written down to the people who participated in the test or questionnaire. However, no one ever questions 'why' or 'why not' somebody wrote or ticked a certain box. Such questions are not explored in PISA-type studies (OECD 2015).

The main pay-off for sponsors of such large-scale quantitative studies is that certain inferences can be drawn. For example, when a certain percentage of the total population has participated in the test, it can be reasoned with a certain degree of accuracy what the result would have been had all the people participated. For example, this total can consist of all the students of a certain age in a particular region or of all adults between the ages of 50 and 55. Under certain conditions the result for the sample is representative of the population. Problems ensue when the results of such tests conducted on a sample serve as a basis for conclusions that go beyond interpreting the test results or ticked boxes. When PISA results are being interpreted by the OECD, things always get particularly controversial when the test results serve as a basis for conclusions on the state of the educational system as a whole and are taken as proof thereof. We will not go into detail on this here but simply observe that similar arguments also apply to other influential, large-scale international studies such as TIMSS.

One such comprehensive quantitative empirical study that can support you in your quest to include mathematical modelling into your hands-on, reality-based mathematics education is known in German as DISUM (an abbreviation approximately meaning 'didactical intervention modes for mathematics teaching oriented towards self-regulation and directed by task'). Challenging reality-based modelling tasks have been developed for students in grades 8–10 (aged 14–16 years old). In the course of this work, teachers are supposed to develop a suitable diagnostic-methodological approach to assist their students with independent coursework (https://ivv5hpp.uni-muenster.de/u/sschu_12/pdf/Publikationen/Schukajlow2006_Schueler_Schwierigkeiten.pdf).

When you work with this type of material, whatever the source, questions that might trigger insight are: What is the cognitive potential of such modelling tasks and what key issues do students have to deal with? How can teachers identify these key issues and how can they intervene in an optimum way? How can student-centred lessons on modelling be set up that foster cognition? We assume that you are interested in finding answers to these questions.

Therefore we once again ask you to note down your answers to these research questions based on your experience or appraisal.
What have you written down?

It is useful and informative for us to look at issues and concerns that have arisen in the context of one important and influential project available in German-language countries. The DISUM website[1] has listed the following results based on their research and these results can be used to generate universal guidelines and insights for implementing reality-based mathematics education.

- Modelling tasks that challenge cognitive skills for the classroom, including 'task spaces' for these tasks and an overview of task-specific cognitive impediments (analyses of solution methods).
- Problem-centred (Seven-step) modelling cycle for teaching and research purposes.
- Student-centred (four-step) version of the modelling cycle (step-by-step guide) with explanations tailored to the task.
- Classification of teacher interventions during problem-solving stages.
- Classroom documentation and analysis of modelling tasks by best-practice teachers.
- Didactic classroom sessions on modelling (with detailed 'prompts').
- Customised tests and questionnaires.
- Findings on how students and teachers deal with modelling tasks, such as motivational levels and difficulty of the tasks.

[1]The website is no longer active because Professor Schukajlow has moved from Kassel University to Munster University.

- Psycho-social dynamics of teacher reactions to student work (playing mistakes back to students).
- Importance of motivational feedback.
- Opportunities for individual work in group settings.
- Feasibility of modelling activities with 'Hauptschul' junior high-school students (lower educational standards than Gymnasium-junior high schools).

The DISUM website is no longer available but several English-language papers report on the project (Schukajlow et al. 2012; Krug and Schukajlow 2013; Rellensmann and Schukajlow 2016). Another good source for teachers is the LEMA (Learning and Education in and through Modelling and Applications) project. This EU funded project (2006–09) focused on professional development of mathematics teachers at both Primary and Secondary level. A resource booklet for teachers is available at following link: http://www.lema-project.org/web.lemaproject/web/dvd_2009/english/teacher.html.

Many of these results, including the modelling cycle for teaching (seven-step) and a student-centred modelling cycle (four-step) are available in English-language publications. A Google search using 'mathematical modelling in schools' will yield multiple papers and several websites such as: http://www.nctm.org. The paper 'Mathematical Modeling in the High School Curriculum' by Hernández et al. (2016), shows how teachers can teach school students mathematical modelling and how to evaluate their engagement. Additionally, a recent report, jointly published in 2016 by the Consortium of Mathematics and Its Applications (COMAP) and the Society for Industrial and Applied Mathematics (SIAM), entitled "Guidelines for Assessment and Instruction in Mathematical Modeling Education" (GAIMME) offers further guidelines in relation to the assessment and teaching of reality-based mathematics (available at http://www.siam.org/reports/gaimme.php). Hopefully, using the all results listed above, we have been able to get you interested in various approaches to teaching and assessing modelling in schools.

A paper written by Werner Blum and Rita Borromeo Ferri (both members of the ISTRON Project) captures students thinking processes as they engage with modelling tasks. In the paper entitled "Mathematical Modelling: Can It Be Taught and Learnt?" the authors discuss the mathematical thinking of the students as they attempted the following modelling task:

> In the bay of Bremen, directly on the coast, a lighthouse called 'Roter Sand' was built in 1884, measuring 30.7 m in height. Its beacon was meant to warn ships that they were approaching the coast. How far, approximately, was a ship from the coast when it saw the lighthouse for the first time? Explain your solution.
>
> (Blum and Borromeo Ferri 2009, p. 49)

Using a simplified model (i.e. plane Earth/sea) and the Pythagorean theorem, you should find a distance of approximately 20 km.

In the paper, the researchers provide insights into the way students approached the problem. They report dialogue from the students and visually represent how students negotiated the modelling cycle as they engaged with the task. If you wish to learn more about this study we direct you to the actual paper which is available at http://gorila.furb.br/ojs/index.php/modelling/article/download/1620/1087.

From our point of view, we would like to make two remarks in relation to this paper: (1) It is always useful to get detailed information about students' thinking and learning; (2) We never find information regarding unmotivated students in research projects like this. Even if they do not reach a conclusion (like Sebastian in the above paper) we find the students are still highly motivated to find an acceptable solution. The authors can confirm the presence of these high levels of motivation having discussed it with the researchers.

8.2 Another Example of Qualitative Research

Qualitative research attempts to find out more in-depth information about small groups of subjects than questionnaires can ever provide. It employs different methods such as interviews, video observation and focus groups. In this approach, there is no attempt to find a representative sample in the first place. When such qualitative research is successful, it provides in-depth insight into special situations and opens up the way to further research.

Let us take an actual thesis that is currently in progress in a European university as an example. Researcher A has done research on the attitudes and emotions of mathematics teachers who have previously studied in his university using detailed interviews. It is informative to discuss how the research is structured.

Researcher A aims to provide a comprehensive overview of the attitudes of nine pre-service mathematics teachers. As part of this study, when investigating what influences the quality of a mathematics lesson, he found two prominent themes: (a) Teachers' beliefs, (b) Teachers' emotions. He went on to state that:

> These beliefs are conscious or subconscious mindsets or philosophies that a teacher has of mathematics itself, as well as mathematics in schools and of teaching and learning mathematics. These beliefs can greatly influence the classes the teacher holds. Emotions … are omnipresent in classrooms in manifold forms.

Furthermore, Researcher A discussed teachers' motivation to perform. He ascertains that emotions stimulated by particular events or situations are generally short-lived and as such tend to have an immediate and short-term effect. However, he discovered that if these emotions continue to be provoked across a range of different scenarios they could influence future, long-term behaviour to a certain degree. In order to gain further insights into the role that teachers' beliefs and emotions play in relation to effective teaching, Researcher A compiled a questionnaire for the nine subjects in his study, consisting of open questions. He also conducted interviews with each teacher.

This overview of the author's project shows a typical set-up for a qualitative empirical study. Picking up on previous research (beliefs and emotions in this case) a specific problem or a research hypothesis is developed and then put to the test. If you are not really familiar with this type of research, you might pose the following question, as many people influenced by mathematical statistics do: As only 9 teachers were questioned, how can the result be representative of a larger group of teachers, let alone all mathematics teachers? It cannot be! However, qualitative researchers focus on getting a better understanding of a human situation or phenomenon by exploring the totality of the situation in depth, using small samples and non-numerical data.

Why do researchers then bother at all with this type of research when it does not produce any representative results? There are good reasons for doing so. For example, qualitative research studies often give us insights pointing the way ahead. Researchers find out things interviewees just would not write on a questionnaire. Without qualitative research methods, these things would go unnoticed. In many cases, qualitative research prompts quantitative studies with a new area of focus and conversely, questionnaires or tests often provide hypotheses for qualitative studies.

The results of the qualitative study mentioned above were similar to those of a larger study by another experienced researcher at the same university (Researcher B). They both focused on types of mathematics teachers, the 'Learning-Process-Teacher' and the 'Transmission-Teacher'. Researcher A sums up his findings on beliefs as follows:

> Let me point out that I could reconfirm the teacher types reconstructed by Researcher B in my research as well. The distinction between these two basic types was entirely unproblematic, as the differences were quite vast. The decision for or against constructivist instruction has many consequences that can easily be determined. This holds true even if a teacher sometimes embellishes his statements a little to be in line with current didactic research on good classroom instruction. From my experience, it becomes apparent what a teacher thinks of student-centred instruction when asked about group work. Is he convinced that his students are able to do meaningful independent work on their own – or does he consider it a waste of time?
>
> (p. 168)

Some more findings to emerge from this study in relation to teacher emotions are:

- The research confirmed theoretical assumptions on performance-related emotions,
- Teachers associated pleasure with student progress,
- Pleasure was also associated with affective-motivational and disciplinary goals,
- Lack of motivation on the part of students was a cause of irritation for teachers,
- Anxiety among teachers is often triggered by new and unfamiliar situations,
- Varying levels of appraisal and confidence in one's own efficacy triggers certain emotions for teachers.

Can you identify with one of the teacher types described?
Do the considerations make sense to you?
Would you like to find out more about the results of empirical research on hands-on, reality-based mathematics instruction?

If so, then let us turn this into a research task for you. Use a Google search based on 'Mathematical modelling in schools' to identify important researchers such as W. Blum, R. Borromeo Ferri, G. Kaiser, R. Levy, M. Swan, and Z. Usiskin. You will find many papers of interest and further references. You will also find various important websites such as the National Council of Teachers of Mathematics (U.S.) (http://www.nctm.org); PLUS magazine (U.K.) (https://plus.maths.org/content/teacher-package-mathematical-modelling); and NRICH (U.K.) (http://nrich.maths.org).

Finally, here is a hint for German-language speakers. A visit to the ISTRON website will yield similar results. There you will find names of researchers such as W. Blum, R. Borromeo Ferri, G. Kaiser, K. Maaß, S. Schukajlow and many more as well as titles or abstracts of research results that have been presented at ISTRON conferences. Among these you can select publications you would like to find out more about.

References

Blum, W., & Borromeo Ferri, R. (2009). Mathematical modelling: Can it be taught and learnt? *Journal of Mathematical Modelling and Application, 1*(1), 45–58.

Consortium for Mathematics and its Applications (COMAP) & Society for Industrial and Applied Mathematics (SIAM). (2016). *Guidelines for Assessment and Instruction in Mathematical Modeling Education (GAIMME)*. Retrieved November 28, 2017, from http://www.siam.org/reports/gaimme.php.

Hernandez, M. L., Levy, R., Felton-Koestler, M. D., & Zbiek, R. M. (2016). Mathematical modelling in the High School Curriculum. *Mathematics Teacher, 110*(5), 336–342.

Krug, A., & Schukajlow, S. (2013). Problems with and without connection to reality and students' task-specific interest. In A. M. Lindmeier & A. Heinze (Eds.), *Proceedings of the 37th Conference of the International Group for the Psychology of Mathematics Education* (Vol. 3, pp. 209–216). Kiel, Germany: PME.

Organisation for Economic Co-operation and Development. (2015). *PISA 2015: Technical Standards*. Retrieved April 25, 2018, from https://www.oecd.org/pisa/pisaproducts/PISA-2015-Technical-Standards.pdf.

Rellensmann, J., & Schukajlow, S. (2016). Are mathematical problems boring? Boredom while solving problems with and without a connection to reality from students' and pre-service teachers' prospective. In C. Csíkos, A. Rausch, & J. Szitanyi (Eds.), *Proceedings of the 40th Conference of the International Group for the Psychology of Mathematics Education* (Vol. 4, pp. 131–138). Szeged, Hungary: PME.

Schukajlow, S., Leiss, D., Pekrun, R., Blum, W., Müller, M., & Messner, R. (2012). Teaching methods for modelling problems and students' task-specific enjoyment, value, interest and self-efficacy expectations. *Educational Studies in Mathematics, 79*(2), 215–237.

Chapter 9
Teaching Real World Mathematics: Some Background Theory

What do you think: do you need to know the basic theory or not when you teach reality-based mathematics in your classes? Studying mathematics at university level gives a strong message that it is absolutely necessary to understand the basic theory before you can proceed with your study. For example, you know that you can't read or understand a proof in Analysis or Algebra if you are not able to handle sets and understand the rules of logic. However, in everyday life you do most things without really understanding them. You clean dishes without understanding all the chemical reactions happening in this process. You use a lot of machines like your TV, PC, car, smartphone and so on without really understanding them. Just for fun, imagine a world where it is forbidden to use a machine you did not construct and build yourself!

Now we ask you to think about the next question that arises: What type of activity is learning to teach reality-based mathematics? Is it like studying mathematics or like learning to solve everyday problems? The connection between the two parts of the learning/teaching activity, namely, *real world* and *mathematics*, opens the door to a two-part answer. The *real world* sounds like everyday life and *mathematics* reminds you of your study.

What do you think now? Our thinking led to what you are holding in your hand: this book! As you have seen as you worked your way through the book our approach is to start with a motivating example, offer ways to get started in the classroom before moving on to more challenging activities. Finally, at the end of this book we think that you might be interested in reading a little about the theoretical background that underpins our approach. There are several books about this theory and this makes our job here quite easy. In this chapter we will give a short overview highlighting three theoretical components of reality-based mathematics education. We acknowledge the work of forerunners in reality-based mathematics education like Freudenthal, the founder of the highly influential Realistic Mathematics Education (RME) movement in The Netherlands in the 1970s, and others. However, it is not our purpose here to write a book on the development of the theory of teaching real world problems but rather to offer a practical guide for teachers and their students.

J. Maaß et al., *Mathematical Modelling for Teachers*, Springer Texts in Education, https://doi.org/10.1007/978-3-030-00431-6_9

We start by *Thinking about Modelling*. Modelling real world problems implies making a lot of philosophical decisions about reality, recognition, mindset, ways of thinking and so on. The second point brings a special focus to this activity by asking you to look at reality with *Mathematical Eyes*. Thinking in this way opens our eyes to connections between mathematics and reality. The third point requires us to take a new look at the relationship between mathematics and reality. We talk about the thesis that mathematics is a hidden source of all so called "new technologies". If mathematical modelling is really successful the new possibilities created are *Mathematical Technologies*, useful devices that work by pressing some buttons without requiring any knowledge about all the very difficult mathematics behind the screen (of a TV, for example).

9.1 Some Thoughts on Modelling

Now we are going to explain some of the theoretical background of (mathematical) modelling. We have refrained from presenting these considerations in more detail thus far, as many other books have been published on the theoretical background of mathematical modelling. What else does modelling entail in our view? We started with the philosophical basics behind modelling as a means of better understanding and changing the world. Whoever wants to set up a mathematical model of a real world situation should also acknowledge some of the philosophical considerations of what 'reality' actually is, and how (and to what extent) it may be perceived. Applicable mathematics education makes some demands on reality and cognition.

There is another important reference to philosophy that has to be acknowledged: everything we do, or do not do, has consequences. What implications does this have for mathematics education? What will our response be, after modelling and changing the world (even if only a little), to the question of responsibility and possible consequences? The question of responsibility immediately takes us to ethics! We flag these issues but they will not be discussed further here.

Another important part of the theoretical background concerns itself with creating a historical record of how modelling influenced mathematics. Part of the meaning of mathematical modelling and applicable mathematics education in schools derives from this. We also need to go into detail about certain theoretical aspects of modelling, such as how to deal with accuracy, the use of computers, study aids that help us solve problems and teaching methods suited to this type of class. We shall end this book with a short theoretical reflection on the modelling process which discusses some of the issues discussed to date (see Fig. 9.1).

At first glance, this diagram does not appear to contain any mathematics. However, the message this figure conveys is that all of us are constantly and naturally using mathematical models in dealing with what we call reality. The second key message is the interdependency between humans, reality and models. Humans create models,

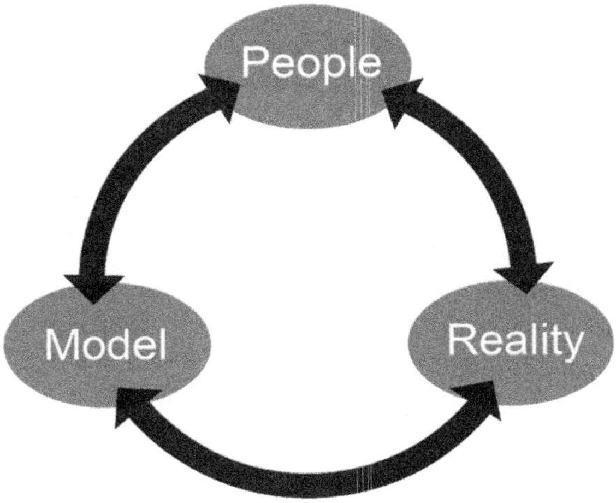

Fig. 9.1 The modelling cycle (We acknowledge that there are a number of modelling cycles proposed in the literature for pedagogical purposes. We have discussed some examples of these, such as the modelling cycle for teaching (seven-step) and a student-centred modelling cycle (four-step), in Chap. 8.)

and through these models they alter reality. Obviously, reality in turn influences humans and the models which are supposed to describe it.

Let us first present you with a few examples. When we go shopping we apply models of objects, people, and behaviours. For example, we plan our route to the shop or we write out a shopping list. If our destination is the bakery around the corner, we do not need to think too hard about how to get there. The route already exists in our mind—as a model. Of course we do not have the actual path (comprising matter like the entire footpath made with tons of rocks) in our heads, but we do have a mental representation or recollection of the way it was when we went there last.

If we want to go to the new shopping mall on the outskirts of town, we might need a plan more reminiscent of a model, for example a map or a public transport map, so that we can figure out how to get there. If we need to purchase new trousers there, we may already have a model of the trousers in our heads, an image taken from a commercial, or a human model who has worn these trousers during a fashion advertisement on TV.

Secondly, we suggest that the way we perceive reality (or whatever we take it to mean) is greatly influenced by pre-existing models. When we go for a walk and see plants and animals, we recognise them as deciduous trees, hedges, flowers, birds, insects, and so on by reverting back to our existing knowledge of plants and animals. In other words we use the knowledge we already have about biology and gardening (which may or may not be accurate), to assign the plant or animal that we see before us to a certain group. Even when we watch children playing football, we employ models belonging to the natural sciences to help us make predictions: will the child score a goal with this kick? Will the child reach the ball before it falls into the brook or rolls onto the street?

Thirdly, we would like to remind you that this aspect of our reasoning is not new to philosophers; we did not invent it. In 1934 Sir Karl Popper already wrote in his *Logic of Scientific Discovery*: "Our ordinary language is full of theories, that observation is always observation in the light of theories" (p. 37). Theories are models of objects, behaviours, organisations and so on, hence this is not a new phenomenon.

You may ask yourself 'why is there an arrow in our diagram (Fig. 9.1) pointing in the opposite direction from models to people?' Obviously, models also influence people as they make it possible for us to perceive something correctly. Someone who has a model of the way a car works in their mind, or a computer, or a virus, can use this image to make better use of their car or computer or to better protect themselves from a virus. Medicine is an excellent example of how visual models of organs, bacteria or genes have led to progress in diagnosis and therapy. Similar models may also contribute towards people living a healthier life.

Models also influence our everyday lives and society. For example, macroeconomic models help justify government tax hikes or tax cuts, or climate models point us to changes caused by the way we humans use energy. The topic of energy consumption by humans reminds us that one element of our diagram (Fig. 9.1) is the word *people*. However, this should not imply that all people are the same, that they have the same perception of reality or interest in change, or chances of wielding influence. This point is reinforced by the following observations. Those who live in a house with central heating and air-conditioning use much more energy than homeless people living on the streets. Additionally, someone who trades in commodities at the Stock Exchange has a different view of corn than someone who cannot feed their family as a result of high corn prices. Finally, someone who uses their bicycle to get to their holiday destination uses up much less oil than someone who flies halfway around the world during their holiday. Similarly, people have a different view of reality and their perception of reality. This depends on the models of aspects of reality that are retrievable in their minds and ready for use. We are certainly aware that from a philosophical point of view what we have presented here is a very simple way of using the term reality, but we do not want to delve any deeper into this from an epistemological viewpoint at this time.

What this simplifies down to is that Fig. 9.1 is a very simplified model of the way humans deal with reality, which always involves modelling. Entire libraries of knowledge, experiences and theories are available on these three terms, but we will not go into further detail on this here.

Our final remark on Fig. 9.1 is that obviously there is a correlation between reality and people. On the one hand there is a framework of social, economic, ecological and personal reality within which humans can move around, while on the other hand, it is humans who influence this framework and thus reality by trying to assess and improve their situation.

And where is mathematics contained in this? Modelling with mathematical methods is a given in research. Research reports from the natural and social sciences and other areas of science naturally contain mathematical formulae (some call these formulae 'laws', hoping that nature and society may adhere to them). In order to justify their results and the correctness of the research methods employed, they refer to the mathematics used. When we looked at mathematics-in-industry in this book, we got the impression that mathematical modelling opens new perspectives on aspects of reality for a deeper understanding and potential opportunities for change. This holds particularly true of topics pertaining to the natural sciences, technology and ecology. However, if individual human behaviour or psychological factors are to be modelled, the limitations of meaningful mathematical modelling soon become apparent. In the everyday-lives of most people explicit models using mathematical methods are rather rare. We have concluded that this is the reason why many people miss out on their chance to arrive at their decisions in a more rational way, or to better understand social, economic or environmental developments. This is all the more reason for us to stress the need for applicable mathematics education in schools.

We will now expand on the 'model' component of our graph to better illustrate the role mathematics can play in modelling.

In Fig. 9.2 we have added more detail to the lower part of our diagram, just as if we looked at the 'model' component through a magnifying glass. Our aim was to set up a mathematical model, so we filtered the data and structure belonging to those aspects of reality that we wished to identify or change. Please bear in mind that this is not a top-down structure in which a human executing some plan models reality according to his will and puts it into mathematical terms—even if some people would prefer that approach. In fact, the correlations are manifold. Already when you select the aspects of reality to be taken into account, you have to determine which ones can be described in mathematical terms. Emotions and other psychological aspects have to be left out of the picture as well as human relations and esoteric concepts. Ultimately, the scope and quality of our mathematical knowledge determines what is appropriate to put into mathematical terms. For example, someone who has never heard of numerical solutions for complex systems of differential equations used to describe currents will most likely not make an attempt at optimising the bow of a ship.

Figure 9.2 is supposed to illustrate the dynamics of a model, as the image contains several cycles (see smaller circle in the lower right section of the model). The mathematical model in the lower half of the figure, is 'only' a special part of the model, an intrinsic and particularly effective method of modelling. The boxes *improved model or new goals* and *altered reality* should remind us that (mathematical) modelling can prompt some changes, and that these in turn may, and should, influence humans. One possible effect of a research-related model could be new insights or questions that may lead to new models.

After the mathematical tasks have been performed for our first model, additional questions often ensue and we try to improve our model by using more accurate data, more information on the structure of the data, or by ignoring certain aspects of our original model. Frequently, we also look for better mathematical models. There is no rule as to how many cycles of the modelling process we need to undertake before we are satisfied with the result or before we abandon the model due to lack of time, lack of methods for further improving the model, or for other reasons. Last but not least we would like to draw your attention to the fact that computers, and the possibilities and limitations of mathematical software, greatly influence the topics and the aspects of the model we choose to focus on, or our chances of arriving at a suitable result. Many topics in research, industry, and school can only be dealt with if sufficient computer support is available.

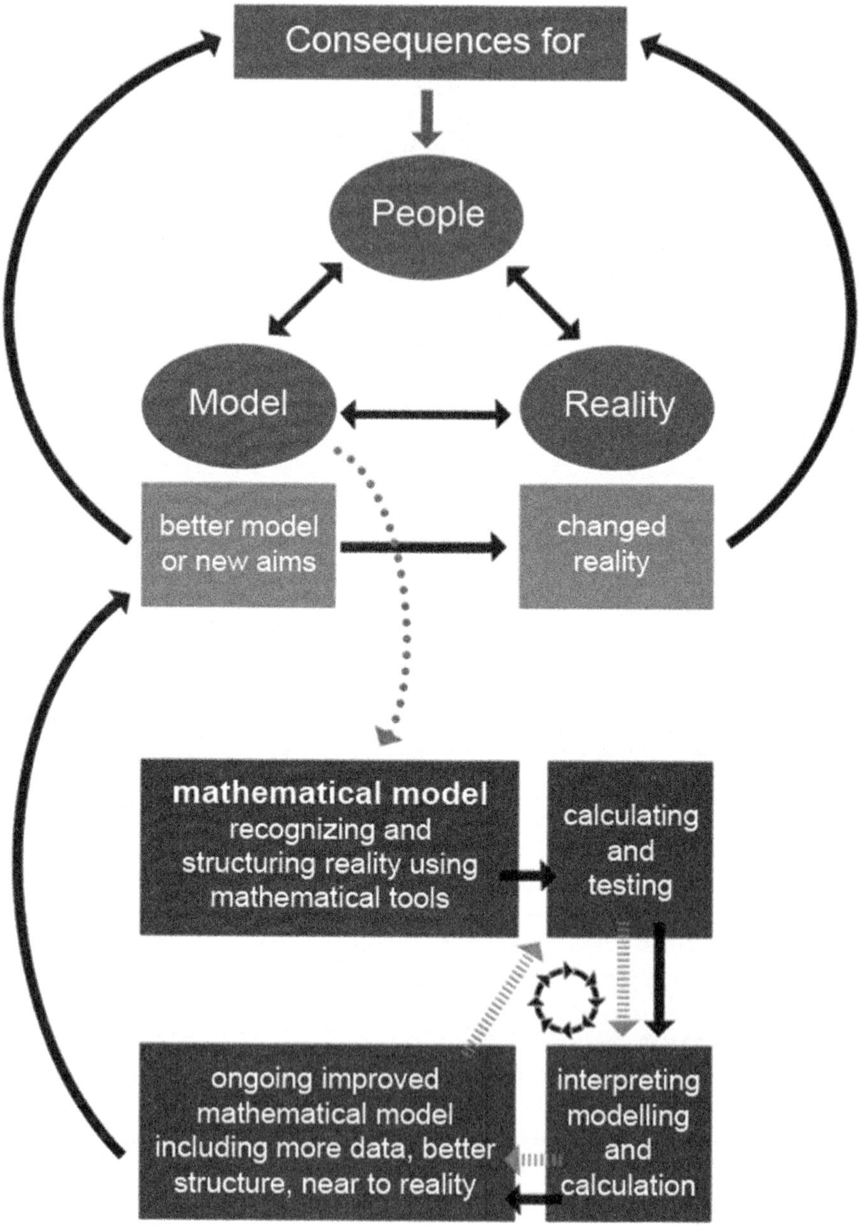

Fig. 9.2 Detailed modelling cycle

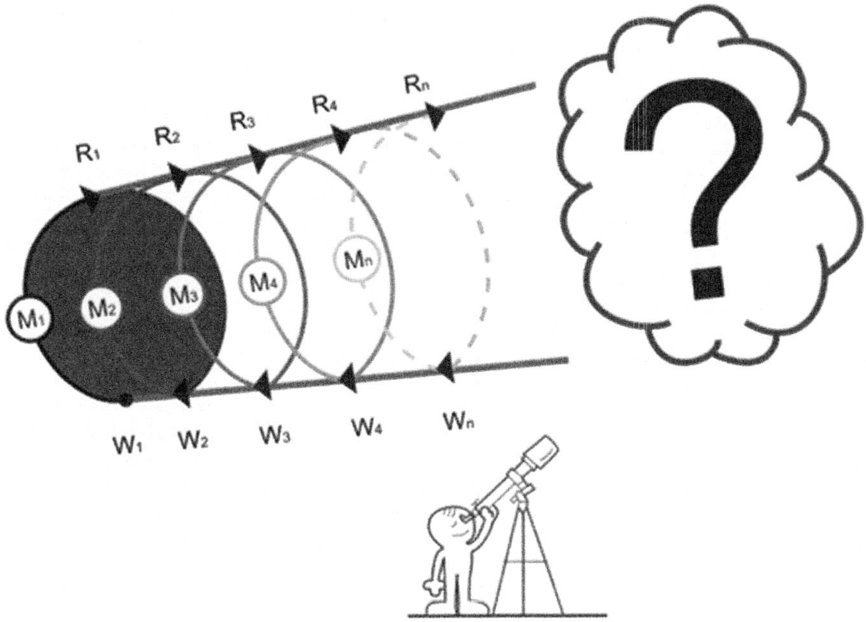

Fig. 9.3 Modelling cycle—in perspective

The third and final diagram (Fig. 9.3) is supposed to illustrate that modelling is an open-ended process, and as such the end cannot be foreseen.

Once again we start with those of us (We = W_1) who wish to comprehend or change something. We set up a model (M_1) and thus provoke consequences that affect reality (R_1). These consequences, or other motivating factors, lead (us – W_2) to renewed efforts and new, hopefully improved, models (M_2) which in turn affect reality (R_2). When, and in what way, we (W_n) end our efforts (at model M_n) at improving our understanding of reality and reality itself (R_n—the last of the changed reality steps), cannot be foreseen at the outset. Nor can our final assessment of the process be foreseen: could we actually (or in someone else's view) achieve a better model than the one we currently have?

9.2 Personal World View and Applicable Mathematics Education

Our aim in writing this book is to produce a vehicle for change in mathematics education. We have invited you as mathematics teachers to change your classroom practice in favour of applicable mathematics education. This involves more than introducing new approaches, it also involves developing a new way of thinking about mathematics education. All of us, whether we know it consciously or not, have a particular view of the world, and in particular mathematics, that guides our teaching and learning. This is our world view. This necessarily involves looking at the world from a specific standpoint using a specific framework of ideas, attitudes and approaches. The authors have invited you to develop your world view in a particular way. In the following paragraphs this approach is made explicit by using two examples viz. Maths Eyes and Mathematics as Technology, rather than pursuing a deep philosophical approach.

9.2.1 Seeing the World Through Maths Eyes

If somebody perceives the world through Maths Eyes, this can influence the way he/she relates to mathematics in two ways. One way would be that he/she recognises mathematical objects in the world around him/her, such as triangles or squares or other geometrical objects. Numerous technical products as well as plants and animals consist of such shapes. A second way involves noticing eye-catching forms, such as a pattern in an object or the structure of a plant that may attract attention and generate interest. Is this crystal only made up of hexagonal structures? How come? The close link between the natural sciences and technology soon leads to the following question: Can we make a crystal grow in a way so that it forms the structure we want it to, possibly a chip with integrated semiconductors? Does this enable us to grow an industrial diamond? Seeing the world through Maths Eyes is not only supposed to provide us with an interesting perspective of the world and to open the door to many useful technologies, it is also about having fun. Photos of children and teenagers who participated in Maths Eyes competitions show students who appeared to really enjoy themselves dealing with mathematics in this unconventional way. Unfortunately, having fun in mathematics class does not occur as frequently as we would wish it to.

When we consider how we perceive the world the answer is obvious—with our eyes! This, however, is only one part of the picture. Our eyes take in optical input (such as light/dark, colour or brilliance) and then transform it into signals that reach our brain via our optical nerves. This is where they will be analysed. The act of seeing, that is interpreting optical information taken in by our eyes, takes place in our brain. How does it actually work? A few rather simple considerations help us understand this very complex process, which is still being researched. Initially the brain forms clusters of individual data, that is optical stimuli that have been converted into neural stimuli, and then correlates them with already existing patterns that have been memorised. This is a ball, a human being, a house, a car, and so on. If a triangle, a spiral or another mathematical object is identified, then an archetype, a sample that it can be compared to, must already exist within the brain. Then the brain filters ‘irrelevant’ data. What exactly is ‘irrelevant’? This may be a parked car (=not dangerous), a tree that is not in our way, or an unrelated person among many others on the other side of the street (=insignificant). Another cluster of data focuses on ‘importance’, but this happens subconsciously. One example would be optical stimuli concerning the path we are about to walk. Is it flat and even? Are there any steps or potholes? Is there anything in the way? Our brain processes this kind of data, and then instructs our muscles and tendons accordingly. Luckily the subconscious takes care of this coordination—otherwise our consciousness and attention would be on constant overload.

In a pedestrian zone or in a shopping centre, we frequently move around without thinking about not stepping on anyone's toes or not bumping into someone. This subconscious, perfunctory processing of data may also be acquired or practiced, as is the case with skiing or driving a car. An experienced driver stays within their lane, sees some braking lights light up and brakes without consciously thinking about it. Similarly, the brain may translate an acoustic stimulus resulting from a certain engine speed into the message 'this speed is ok' (=nothing needs to be changed) or 'it is too loud' (=I need to take a look at the speedometer).

The data that can pass all these filters and enter our consciousness is of paramount importance in order to learn and apply something we already learnt. Interest and motivation form an additional filter or turnstile here. We hardly notice things we are not interested in, unless we are forced to. A stop sign on the side of the road is something we must acknowledge, even if we do not feel like it. When learning something, motivation is the determining factor for our interest and our attention filters. If it is only fear of getting a bad grade that motivates us (something extrinsic) to pay attention in class, the likelihood of satisfying the learning outcomes will be worse than if our motivation comes from within (intrinsic), that is, if we really want to know about and learn something. Certainly, our progress will be faster and more permanent, if it is intrinsic motivation that drives us to want to learn something.

It is exactly this kind of intrinsic interest in mathematics that the aforementioned competition was about. Those of us who intentionally tune our attention to the mathematics that may be discovered in the world around us suddenly see the world with different eyes—a new perspective. Those of us who actually start to perceive mathematics everywhere, do not need to ask: Why do we have to study this? It is obvious that mathematics is present everywhere. Once we realise this it becomes obvious that it may be very useful to study and understand mathematics.

Understanding the World Better and Changing it Through Mathematics

The brain will recognise mathematics in data perceived by the eyes, if it already knows something about mathematics. When looking at a honeycomb, we will recognise a hexagon if we already heard about it, perhaps in geometry class. Does this also work the *other* way round? If we take a thoughtful look at a honeycomb, this may arouse our curiosity and lead us to ask: Why do bees build something shaped like this? Would triangles or squares or circles not be easier and better? Detailed mathematical analysis shows that hexagons are perfect in some respects: they use up very little material while forming a stable structure.

Another example taken from our everyday-lives are clothes. Those of you who have already attempted to sew a simple garment from a panel of fabric will have realised that it is not all that simple. Which pattern must be cut from a panel (of fabric) so that it will turn into a three-dimensional skirt? In mathematical terms this is a truncated cone. When we cut up fabric, mathematics is not applied; sewing patterns from magazines are used instead. But who creates complicated sewing patterns for evening gowns and traditional clothing, such as alpine fashion? As is often the case with handicrafts and technology, mathematics is applied in a hidden way; it is used without explicitly stating it and so many people are not aware of it. Those of you who do not just wish to copy and apply things will have huge successes by applying mathematics deliberately. After all, all new technologies are essentially mathematical technologies.

9.2.2 Mathematics as Technology

We will close this chapter with a brief digression that shall demonstrate how the approach to, and use of, mathematics presented in this chapter reflects on the wide field of mathematics and its image.

During the long history of mathematics there has been a coexistence of abstract and applied mathematics, since the height of the ancient Greek civilization some 2,500 years ago. One important aspect of the history of mathematics was the development of theories, of schools of thought and logic as seen in geometrical proofs. Yet at the same time the use of mathematics for professional and everyday life, for example for land surveying or planning for manual crafts, was equally important. The theoretical question whether a certain angle may be divided into three equal parts using a divider and ruler has prompted people to think about this for many centuries. In reality any angle can be divided into three equal parts to an appropriate degree of accuracy.

Up to the 19th century most scholars performed abstract as well as applied mathematics at the same time and often were naturalists too. Extensive educational reforms in the German-language countries, which are often linked with the name of Humboldt, also prompted university research centres for abstract and applied mathematics as we know them today to be set up (such as Göttingen, where Felix Klein and David Hilbert held professorships). Technomathematics (Helmut Neunzert, Kaiserslautern) and mathematics-in-industry (Hans-Jörg Wacker, Linz) were recent additions of the 20th century, which have since been pursued in many other places and met with huge success.

Looking at mathematics as technology is still a different matter than pursuing technomathematics. To this day, it still is a bone of contention whether such a characterisation is admissible or not—after all mathematics is the 'queen' of science to many people. It might help the cause to say that technology is one aspect of mathematics. This implies that mathematics is not reduced to a 'mere' technology, but that technology is one aspect of it. This topic was first raised during an international conference in 1988 and later developed in a journal article by Maaß and Schlöglmann (1988). Now this term has become widely accepted, as a lot of published papers, for example, papers published by the Fraunhofer Institute in Kaiserslautern and Sankt Augustin (http://publica.fraunhofer.de/eprints/urn_nbn_de_0011-n-464240.pdf) proves. This paper also contains plenty of examples showing how many successes have resulted from interpreting mathematics in this way.

9.3 Conclusion

Dear reader, our book ends with this digression on modelling and personal world views. We hope that you, your students and consequently your mathematics classes will benefit from the book and its contents and we very much look forward to your feedback.

References

Maaß, J., & Schlöglmann, W. (1988). The mathematical world in the black box—The significance of the black box as a medium of mathematising. *Cybernetics and Systems, An International Journal, 19*, 295–309. https://doi.org/10.1080/01969728808902171

Popper, K. (2005). *The logic of scientific discovery*. London: Routledge.

Printed by Amazon Italia Logistica S.r.l.
Torrazza Piemonte (TO), Italy

11398557R00103